Farm Animals

The Duchess of Devonshire is the youngest of the seven children of the 2nd Baron Redesdale, and sister of the writers Nancy Mitford, Jessica Mitford and Diana Mosley.

She married Lord Andrew Cavendish, who succeeded to the title on the death of his father in 1950. She works tirelessly at farming and all aspects of the management of the Estate at Chatsworth and it was from her idea in 1973 that the Farmyard at Chatsworth was established. In 1990, 115,000 people were able to see the farm animals there and to learn about farming as it happens today.

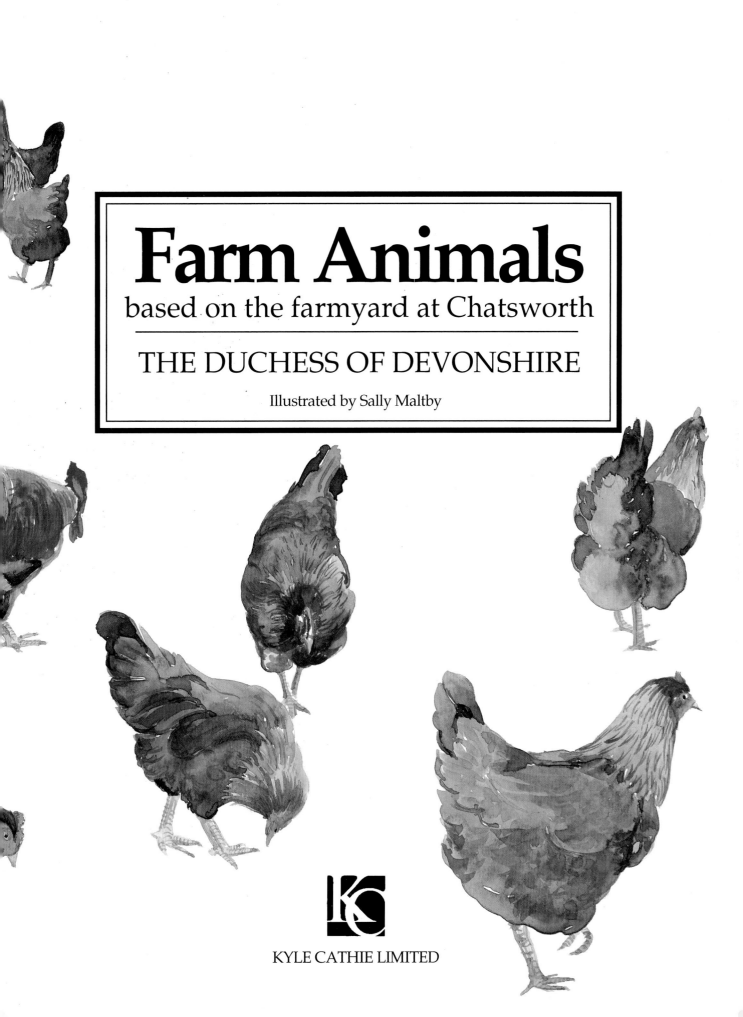

Farm Animals
based on the farmyard at Chatsworth

THE DUCHESS OF DEVONSHIRE

Illustrated by Sally Maltby

KYLE CATHIE LIMITED

by the same author
THE HOUSE – A Portrait of Chatsworth
THE ESTATE – A View from Chatsworth

First published in Great Britain by
Kyle Cathie Limited
3 Vincent Square London SW1P 2LX

ISBN Hardback 1 85626 040 2
ISBN Paperback 1 85626 026 7

A CIP catalogue record for this book is available from the
British Library

House Editor and researcher: Sue Collins
Designed by Geoff Hayes
Printed in Belgium by Proost International
Book Production

Acknowledgements
I would like to thank the following people for their kindness and patience in answering
questions and finding facts and figures which were a great help to me:
   The goat experts – John Wood-Roberts of the Angora Goat Society; Mrs H Pollock of the
British Cashmere Association; Mrs S Knowles of the British Goat Society; Mrs Christopher
who keeps Angora goats; Mrs Bobbie Clarke, supplier of goat cheese to the Farm Shop; Mr
& Mrs Roberts of the Domestic Fowl Trust who have a wonderful collection of rare
poultry open to the public; Alastair Dymond of the Rare Breeds Survival Trust; John
Ryman a long-time breeder of Large White pigs who introduced me to the Royal
Smithfield Club many years ago; Lorne Nelson, Rupert Turner and Alastair Morrison who
I bothered on the telephone about deer; Sandy Boyd, manager of the Chatsworth Farm
Shop and Sean Feeney, head butcher there; Ian Turner, manager of Chatsworth farms and
Shiela Blackburn, his secretary who is never surprised by the oddest questions; Derrick
Penrose and Roger Wardle, the Chatsworth land agents who know all the answers;
Andrew Middlemiss and Mark Richardson of the Chatsworth Farmyard; Mrs Evelyn Aris-
Fowkes who understands word processors in the Estate Office; Michael Pearman,
librarian and archivist at Chatsworth who can put his hands on the most obscure books of
reference; Tom Askew; Christopher Marler of Flamingo Zoo Gardens and Joe Henson of
Cotswold Farm Park; Dick Norris, retired head keeper, Sean Read, head keeper, and
Derek Neave, keeper, who lent me excellent books on vermin; Alan Shimwell, driver, who
has added photography to his other accomplishments; Helen Marchant, who is my
husband's secretary but somehow manages to do lightning work for me too; Kyle Cathie,
who suggested this book and has published it, and her daughter Josephine who likes dark
brown Welsummer eggs. Thank you all very much.

Photographic credits
Nic Barlow 21, 22, 53, 59, 61; British Chicken Information Service 37; Bruce Coleman Ltd
16, 38; Compassion in World Farming 30, 31,39, 44; Daler-Rowney Ltd 33; Mary Evans
Picture Library 10, 14, 35,; Harringey Meat Traders Ltd 32; Geoff Hayes 11, 17, 28, 30, 34,
36, 44, 48, 52, 53, 55; Jacqui Hurst 15, 26, 27, 49; IPC Magazines Ltd 43; Lyons Maid 14;
Sally Maltby 50; Nature Photographers Ltd 40, 41; PA Belgium 30; Reed Farmers
Publishing Group 14, 17, 23, 24, 36, 62, 63; John Rawlings 19; Rolls-Royce Motor Cars Ltd
19; Alan Shimwell 12, 13, 20, 43, 48; Bobby Tulloch 20, 24, 25; Wales Tourist Board 15;
Whitbread & Co Ltd 55.

The publishers would like to thank Nigel Fortun, Angela and Roger Cherry, George, James
and Jane Gray, John Hardy. Grateful thanks to 'The Field' for permission to quote on p21.

Dedication
**To the memory of George Hambleton and Tommy
Jones who devoted their lives to the cattle, sheep,
horses and ponies at Chatsworth.**

Contents

Introduction

Chatsworth house, garden and park have always been open to the public, so it is a traditional place to spend a day out for the thousands of people who live nearby. Many school parties are among them.

There is a great interest now in how the land is used to produce food and in the conservation of landscape. We got many letters from teachers asking if their pupils could learn about the farms and the crops they produce. The practical difficulties of showing the variety of operations and the limited time of most of our visitors gave me the idea of gathering a cross section of our farming activities into some buildings near to the car park to explain the basic facts of how milk, meat, eggs and wool are produced. The animals you see are not show specimens but are typical of the commercial types with which the farms are stocked. We have added pigs, rabbits and poultry for general interest.

Chatsworth farms include high moorland rising to 400 metres above sea level, 'marginal' land next to the moor which is termed a Less Favoured Area, the grass fields round Pilsley Village and 1200 acres of grazing in the park. There are herds of about 120 Red deer and 240 Fallow deer, a milking herd of 120 Friesians and 4000 ewes. A small flock of about 50 Jacob sheep lives on the Estate golf course.

We keep about 220 Hereford/Friesian suckler cows which are mated to Limousin bulls. They suckle their own calves for 7 months. The calves are then weaned and sold at about a year old as 'stores' to other farmers. The milk from the Friesians is sold to the Milk Marketing Board which supplies Sheffield dairies.

The lambs are sold at our annual sheep sale on the farm on the third Saturday in September. About 5000 to 6000 sheep are auctioned that day. I believe it is the biggest sale from one owner in the country.

Nearly all the land in Great Britain is farmed in one way or another. Even the mountains, hills and moors produce a crop of some kind; the hardier breeds of hill sheep, deer and game birds like grouse, live on heather moorland. The woods, too, are planted, thinned and felled by their owners. Trees are a crop in just the same way as wheat, oats and barley are planted and harvested when ripe, but they take up to 120 years to grow instead of one summer. The country as we know it has been looked after or 'managed' by landowners, farmers and foresters for hundreds of years, and they have kept the fields and woods enclosed and neatly bounded by hedges, fences and walls as we see them today.

We are apt to take all this for granted because we are so used to seeing the views we know. The look of the country would soon change if farmers stopped work and the fields would look like the overgrown, unkempt garden of an empty house.

British farmers produce a large percentage of our food - milk, meat, poultry, eggs, vegetables, potatoes, fruit, wheat to be ground into flour, oats and barley (which are mostly used for cattle food), and other crops like the yellow-flowered oil seed rape and the pale blue-flowered linseed, which go to enrich winter food for cattle and sheep.

Some farmers breed thoroughbred horses for racing, and many keep riding horses and ponies.

The types of soil and weather in these islands vary a lot and so do the crops. Where the rainfall is high – in the West Country, Wales and the Lake District - grass is the most successful crop. Dairy cows need plentiful grass so the farmers here concentrate mostly on producing milk, which, in turn, gives us cream, butter and cheese.

Sheep depend on grass for their food, but they are content with that which grows on the poorer land and so they thrive on the hills where the grass is not so lush. They give us meat and wool.

The rainfall is lighter in the Midlands and eastern counties and the good soil is suitable for growing cereals - the word which includes wheat, oats and barley. Some land is specially suitable for potatoes and the biggest crops are grown in the rich Fenland farms. Kent is renowned for its fruit, so are Devon, Somerset and Hereford & Worcestershire, where much of the apple crop is made into cider.

No two farms are exactly alike and no two fields on one farm are alike. Farmers themselves are the most individual of beings and each has his own way of doing things, even if the end product is much the same. Hence, some fancy one breed of cattle, sheep or pigs, and some another, but all are striving to produce food and make enough profit to earn themselves a living.

Many farmers own the land they work, but some are tenants. This means they pay rent to a landowner who may be an individual or a big organisation like an insurance company, the Church of England or the National Trust. A tenant farmer has the advantage of only needing enough capital to buy the animals and machinery, as the landlord provides the land, the farmhouse and the farm buildings in return for rent.

Because British farmers are so efficient they produce too much food. The EC guarantees to buy the surplus grain and meat and to store it. This is very expensive and politicians are forever trying to find a better way to deal with the problem. Meanwhile many farmers cannot make a living at the prices they presently receive for their produce, so they are looking for other ways to support themselves and their families. Some offer bed and breakfast, some have opened their farms to visitors at a charge, some have opened farm shops, or run a pony-trekking business. Many have started Pick Your Own fruit and vegetables, some go in for deer farming for producing venison. Farmers everywhere are experimenting with new ways of bringing more income - milking sheep, making ice cream and farmhouse cheese, growing herbs, fish farming, 'put and take' fishing, planting vineyards to grow grapes and make wine, beekeeping for honey, breeding rabbits and even snails, and lots of special home-made foods.

Organic food is becoming popular. This means that the land on which the crop is grown or which the livestock graze has not been treated with pesticides or artifical fertiliser for at least three years. Without such chemical aids the land produces smaller crops so the food is more expensive to buy, but many people are prepared to pay more for organically produced food.

For the owner or tenant of a small dairy farm who has no extra help life can be hard and the hours of work long. Cows have to be milked twice a day, so there are no days off or weekends away for him. Calves and lambs are born at any hour of the day or night and have to be looked after, however inconvenient it may be for their owner, who has to be ready to go out in all weathers to care for them.

Farmworkers on the larger holdings are highly skilled people who usually specialise in a certain job. A herdsman milks his cows and rears the calves and young stock (called followers). Like everyone who looks after animals, he or she is a most important person because the health and welfare of the stock demands great responsibility. Some people have the knack of noticing anything wrong with an animal at a very early stage and decide whether to call the vet or whether it is some ordinary ailment which can be dealt with. They must understand about feeding the animals, how much to give them and what suits their farming system best, arrange the matings and act as midwife to the cows when their calves arrive.

The tractor drivers have to be mechanics to deal with their complex and expensive machines. At Chatsworth they harvest the silage and the hay in the summer and haul it to the silage pits and Dutch barns where it is stored till required in the winter. Life can be very frustrating when the weather is bad for these vital jobs and it pours with rain on hay which is drying nicely, so it has to be turned again, thus prolonging the harvest and spoiling the quality of the hay.

When the weather is good they work till it is nearly dark and often right through the weekends. June, July and August, when most people are taking holidays, are their busiest months.

In the winter they feed and clean out the cattle which are in the sheds. As the manure builds up they take it from the sheds and spread it on to the land to fertilise it and ensure a good crop of grass the next summer. They help build new sheds or roads, and repair old ones. They can do lots of jobs on the farm and are always busy.

Shepherds never seem to have a slack time because sheep demand a lot of attention. In spite of looking so hardy and being able to find a living on ground which does not grow good enough grass for other farmstock,

Sheep are prone to many diseases, so they depend on the shepherd for their welfare. He makes sure they are properly fed, and at regular intervals they are dipped in water which has chemicals in it, killing the various parasites which live on sheep. Dipping is required by law.

At lambing time a good shepherd can save the lives of many lambs (and ewes) by helping at difficult births and making sure that the ewes suckle their own lambs by putting them together in a separate pen as soon as they are born. Otherwise another ewe may try to steal the lambs which results in 'mismothering'.

At Chatsworth we have a shed which holds 1000 ewes. It is like a big maternity hospital in April when the lambs are born. They go out in the fields aged about 3 days old. The grazing sheep keep the grass short in the park at Chatsworth which makes it pleasant for people to walk on. A few ewes are put in the Edensor churchyard for the summer where they do a great job in keeping it tidy.

On arable farms the men are responsible for the enormous - and very expensive - machines. Combine harvesters, the tractors which haul the ploughs and the other cultivating machinery mean big investments for the farmer and they must be kept in good working order. So must the grain drying machinery and the buildings where the corn and potatoes are stored after the harvest.

There are many arable farms where no animals are kept, nevertheless the crops they produce are living plants and must be carefully looked after from sowing till harvest, which demands great skill from the farmer and his staff.

There is much to be learned from farmers and farmworkers, and most of them are pleased to answer questions from people who are interested in their many activities.

When in the country it is important to remember that even the grass we walk on is a crop, and that gates must be shut and tins, bottles and plastic bags are not left behind after a picnic. These can be fatal to animals.

All of us at Chatsworth are glad to see people enjoying a day out in beautiful surroundings, and I hope the readers of this book will look at the scenery and the animals with renewed interest.

Deborah Devonshire

The Duchess of Devonshire
January 1991

Dairy Cows

The four main dairy breeds are the Friesian/Holstein, the Ayrshire, the Jersey and the Guernsey. The black and white Friesian/Holstein outnumbers all the other breeds put together, and over 80% of our milk comes from them. Dairy cows are lighter and more feminine looking than beef cattle. A heifer has her first calf when she is about 2½ years old. When she has her second calf she becomes a cow. With good management, she will then have a calf every year until she is 8 – 10 years old. Some breeds, including the Ayrshire, produce calves until well into their teens. The cow, together with the sheep and the goat, was among the earliest of the wild animals to be domesticated, about 9,000 years ago, by men of the New Stone Age. Like the hippopotamus, pig, camel, llama, deer, giraffe and antelope, it is a 'cloven-hoofed' animal with two toes capped with two hooves that look like one hoof with a split in the middle. It is not unusual for a dairy cow to give over 30 litres of milk a day.

THE SANCTITY OF THE COW
The Hindu religion believes that the milking cow is connected with the god and various of their ancient verses refer to the cow as a goddess. Killing a cow considered as sinful as killing a Brahman (a Hindu high priest. *See also Brahman page 17)* and so was strictly forbidden. In India today, killing cows has become a political issue as well as a religious one.

Friesian/Holstein
The Friesian breed originated in Holland. It has flourished in Britain, on the continent of Europe and in the USA, not only because of its high milk yield, but also because it provides good beef calves when they are sired by a bull of a beef breed. Lately Holstein blood has been introduced from America. They too originated in Holland and are the most productive breed of dairy cow.

Jersey
The Jersey and Guernsey, dairy cows from the Channel Islands, are well-known for their wonderfully rich milk, caused by a high fat content. The cream and butter are deep yellow. The Jersey is a small cow and so needs less food than a Friesian/Holstein.

Ayrshire
Named after the Scottish county from which it originates, the Ayrshire is a hardy breed and is common in Scotland and England. Its popularity as a dairy breed has declined slightly because, although its milk is excellent, it does not produce as much as the Friesian/Holstein. Typically for a pure dairy breed the Ayrshire does not have much meat value, as it converts its food into milk, not beef.

DIGESTION

A cow has 4 stomachs. When a cow eats grass she tears it with her tongue and swallows it without chewing it.

This food is stored in the first stomach (the rumen). From time to time she returns this unchewed food or 'cud' to her mouth. It is soaked in saliva and she chews it (chewing the cud or 'ruminating') and swallows it again. The food is then returned to the second stomach (the reticulum) where digestion begins.

In the third part of the stomach (the omasum) surplus water is removed.

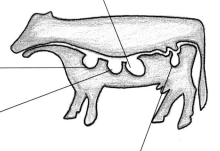

Eventually the food reaches the fourth or true stomach (the abomasum) where gastric juices continue digestion. Any food surplus to the cow's needs is turned into milk.

airy Shorthorn
he Dairy Shorthorn is the traditional English reed of dairy cow. They are excellent milkers nd the bull calves are good for fattening for eef. Now the black and white Friesian/Holstein as overtaken the Dairy Shorthorn as Britain's ost popular dairy cow because they give more ilk than any other dairy breed.

Brown Swiss
The Brown Swiss originated in Switzerland but is now found in North and South America as well as in Europe. Dark brown in colour with a wedge-shaped body and a black nose and tongue, this is a good milker which is also strong and hardy.

ied Rouge
riesians and Simmentals are extremely popular with French farmers, but rance also has its own Pied Rouge (Red Foot). This breed is rusty and hite in colour, and is an excellent milker.

FOOD FOR THE HERD
In summer the cows are turned out into fields to feed on grass and foods called concentrates are fed when they come into the parlour to be milked twice a day. Concentrates are usually made of cereals which give the cow additional energy and protein to keep up their daily yield of milk and to balance their diets.

Hay

Hay – grass that is cut and dried in the field during the summer months when the grass is plentiful – is made into bales of various shapes, and stored in barns until needed in winter.

Silage

Silage is also grass, but it is cut earlier than hay and is not allowed to dry. It is 'pickled' by being kept airtight under or wrapped in polythene. The grass will ferment but the lack of oxygen does not allow it to rot.

Nuts

These are easily handled and are tipped into the trough by machine. In the winter the concentrates fed to the cow are related to her milk yield.

Milking

Before a cow can produce milk she must first have a calf, because in nature the milk is for the calf. She produces the maximum amount of milk for about 6 weeks after the calf is born and stays in milk for about 10 months. During this time - called the lactation period - the cow can give about 6,500 litres of milk, an average of 25 litres a day. Cows are milked twice a day, in the morning and late in the afternoon. This is done in a parlour fitted with automatic milking equipment. The herdsman has a close view of the cows so he can see all is well with them. When in milk, their udders become uncomfortably full after 12 hours and so they queue up readily for their turn to be milked.

WEANING CALVES

When a calf is born the cow produces a special first milk, called colostrum. It is essential for its health that the calf drinks the colostrum because it contains antibodies which protect it from disease. Most farmers leave the calf to suckle (suck milk from its mother's udder) for about 4 days. It is then taken away and taught to drink normal milk, or dried milk mixed with water, from a bucket. When it is 3–5 weeks old, the calf is gradually introduced to solid food, and it is weaned at 5–8 weeks old.

HAND MILKING

The fingers are placed around the teat and the teat is gently pushed upwards.

The fingers are then pressed together, from the forefinger down to the little finger, and the teat is gently pulled down.

The pressure from the fingers squeezes the milk out.

CHATSWORTH PARLOUR

At Chatsworth we have a herring-bone parlour with 16 standings. The herdsman and his assistant milk the herd (see Chatsworth Herd) between 6.15 and 8.30 in the morning and again between 3.45 and 5.45 in the afternoon. Once the milking is finished they thoroughly clean and disinfect the parlour and all the milking equipment.

Each standing in the parlour has a trough at the front. During milking the cows are given concentrates, such as nuts, the amount being carefully measured according to the milk yield of each cow. The food is tipped into a trough from a cone-shaped container called a hopper.

It has been discovered that playing classical music during milking appears to relax the cows. Pop music does not have the same effect!

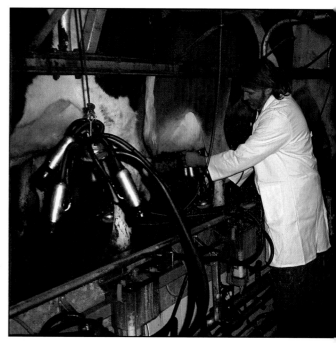

MILKING TIME

The herdsman attaches the clusters from the milking machine to the cow's teats, ready to start the milking process

The milk flows from the teat-cups along a pipe into a container that measures the yield taken from the cow.

MILKING MACHINES

Milking machines are used by almost all dairy farmers today. After washing the udder, the teat-cups or clusters are attached to the cow's teats. The machine is operated by a vacuum pump that draws the milk from the udders through the teat-cups in a sucking action. The teat-cups have rubber liners which pulsate and draw the milk out, imitating a suckling calf.

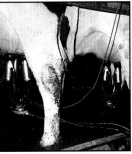

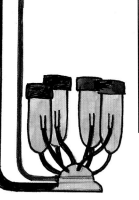

The milk is then directed along another pipeline into a bulk tank – a refrigerated storage tank where it is kept cool for collection.

THE HERRING-BONE PARLOUR

This is the most common sort of milking parlour. Groups of 5–16 cows stand in positions (called standings) on each side of the parlour that are angled in a zig-zag or herring-bone shape. In the centre there is a pit where the herdsman stands. The first group is lined up on one side and their udders are washed, then the cows are milked. While they are being milked the second group is lined up on the other side of the parlour and their udders are washed. As soon as the first group is finished, the teat-cups are transferred to the second, the first lot are let out and another group comes in to be washed, and so on. The cows very soon get used to this procedure and know where to go. It takes about 5 minutes to milk each cow.

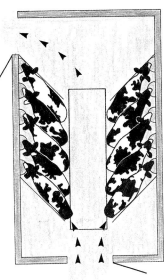

CHATSWORTH HERD

The Chatsworth dairy herd, around 160 cows, are all Friesian/Holsteins. The herd grazes out in the fields during the summer until about late October, when it is taken in for the winter. Some of the milk is taken to the Chatsworth Farm Shop (where it is made into yoghurt, cream and ice-cream) and the rest is collected daily by the Milk Marketing Board tankers. The herdsman, who spent about four years as an assistant, has been in charge of the Chatsworth herd for seven years.

WHAT IS IN MILK

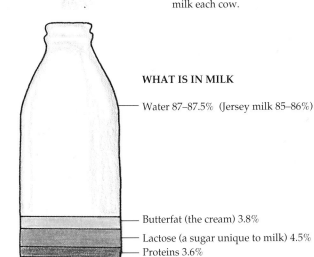

Water 87–87.5% (Jersey milk 85–86%)

Butterfat (the cream) 3.8%

Lactose (a sugar unique to milk) 4.5%

Proteins 3.6%

Minerals and vitamins .75%

From the Cow to your Fridge

Most of the milk produced in this country is sold to the Milk Marketing Board (an organisation owned by the farmers themselves), and they sell it to dairies. At Chatsworth milk is collected by tanker once a day from the farms. It is tested for bacteria, measured for butterfat content, protein and antibiotics. It is then strained and heat-treated either by pasteurisation (*see Louis Pasteur*), ultra heat-treatment (UHT and often called Long Life Milk) or sterilisation. UHT raises the temperature of the milk to 132°C and then cools it. Sterilisation raises the temperature above boiling point (100°C) for a longer period. Homogenisation is a process which breaks up the milk fat globules so that milk and cream are mixed and the cream does not rise to the top of the milk.

Nearly half the milk produced is sold as liquid milk. The rest is made into butter, cream, dried milk and yoghurt. Britain is unusual in having milk delivered daily to the doorstep, and our milkmen deliver 20 million bottles every day.

TANKER
A tanker collects the milk from the previous afternoon's and that morning's milking from each farm. Before the driver takes the milk from the bulk cooling tank, he checks that it is fresh and agitates it to mix the cream with the rest of the milk. He measures the amount taken and removes a sample that will be tested at a laboratory. The temperature is 4°C when it leaves the farm. The tanker is insulated s[o] that it keeps the milk cool like a huge vacuum flask while it is being transported. It can carry 10,000 litres.

LOUIS PASTEUR (1822 - 1895)
Pasteur was a Frenchman who discovered that bacteria were the cause of milk going sour, and that if the bacteria were heat-treated they could be killed. The hotter the treatment the longer the milk keeps; short heat-treatment means less change in colour, flavour and nutritional value.

MILK BOTTLE TOPS
The different types of milk can be identified by the colour of their foil caps.

Untreated	Pasteurised	Sterilised	Homogenised

Channel Islands (untreated)	Channel Islands (pasteurised)	Semi-skimmed	Skimmed

MILK PACKAGING
Bottles, waxed cartons and plastic bottles are used. High-speed machines fill and cap around 800 bottles a minute.

ICE-CREAM
Ice-cream came simply from freezing cream. Now it is made from either cream or custard, or a mixture of both, and mixed with fruit or nuts.

BUTTER

Cream skimmed from the milk is made into butter. About 10 litres of milk produce about half a kilogram of butter.

CREAM

The creaminess of milk depends on the amount of butterfat it contains. Cream rises to the surface of milk because it is lighter. Much of the cream produced in this country is turned into ice-cream. Cream and butter from the milk of Channel Island cows (Jersey and Guernsey) is easily recognised by its deep yellow colour.

YOGHURT

The nomadic tribes of the Middle East made yoghurt thousands of years ago. Bacteria (known as a starter) are added to fresh milk and incubated, so that the bacteria reproduce themselves and the mixture thickens. For commercial use today, yoghurt is homogenised and then pasteurised. Yoghurt can be made from skimmed milk or from whole milk.

CHEESES

Hard cheese

This is the oldest way of preserving milk. The milk is pasteurised, cooled and pumped into vats. There it is treated with a starter called rennet; this comes from the digestive juices of a calf, and the effect of the rennet causes the milk to set. It is then heated, stirred and cut so that it separates into the solid curds and the liquid whey. Salt is added to the curds, which are then put into moulds and pressed. Traditional cheeses (as opposed to the supermarket type of block cheese) are ripened in cool rooms and sold anything from six months to three years later.

Cream cheese

We have a huge variety of soft cheeses now - home produced ones include Lonchester, Belvoir Blue and many varieties of locally made curd cheeses.

CHEESE MAKING

Processing the curds on a Welsh farm.

1 Roule
2 Huntsman
3 Stilton
4 Edam
5 Sage Derby
6 Pont l'Évêque
7 Brie
8 Caerphilly
9 Cambozola
10 Emmental
11 Red Leicester
12 Cheddar
13 Red Windsor
14 Shropshire Blue
15 Cottage Cheese

Beef Cattle

Most of the beef we eat in this country is produced by fattening the male calves (bullocks or steers) from a dairy herd. But there are specific beef breeds: Beef Shorthorn, Hereford, Aberdeen Angus, Devon, Sussex, Highland, Welsh Black, Galloway and Lincoln Red. Hereford and Aberdeen Angus produce the finest quality beef. Among European imported breeds which give lean meat Charolais and Limousin (French) and Simmental (originally Swiss) are the most popular to be crossed with dairy cows. On some farms they are pure-bred or crossed with other beef breeds. The bull calves are nearly all castrated, which means that they are unable to reproduce.

Hereford

A pure beef breed associated with its native county. Its white face comes from crossing original red breeds with white-faced Flemish cattle. It is exported throughout the world. It does well on grass and can survive in hot dry countries better than some other cattle. When the Hereford bull is crossed with other breeds, the calves have the distinctive white face of their Hereford sire.

RINGING

At about 10 months a bull kept for breeding has a ring put in his nose. Bulls are extremely strong and heavy and are of uncertain temper. A rope slipped through the ring enables the handler to control him.

An early wall-carving of a cow in the Temple of Horus, Edfou, Egypt.

Aberdeen Angus

The Aberdeen Angus is one of the best native beef breeds. As both a pure-bred and a crossed animal, it has won awards at stock shows throughout Britain and America. The Aberdeen Angus was first introduced to America in 1873 and has been widely popular since. This breed is a good example of a 'polled' animal; naturally it has no horns.

DEHORNING

Cattle, sheep, goats and antelopes have horns, which grow from the animal's head. They differ from a deer's antlers in that they are not all bone. The middle section is bone, and the rest is horn. Calves are de-horned at a few days old which prevents them injuring each other when they are older. This is done under local anaesthetic. Some breeds have been selectively bred so that they are born without horns. Cattle and sheep that are naturally hornless are described as 'polled'.

Brahman

In hot dry areas infested with ticks, such as the states of Texas and Colorado in the USA and some countries of South America, it was desirable to produce a breed that could tolerate heat and insect attack. The Brahman cattle, which originated in India (*see The Sanctity of the Cow, page 10*), were imported into the USA in the early 1900s, and were crossed with other breeds. The Beefmaster and the Droughtmaster are the results of two of the most succesful crosses. The Brahman has a hump on its shoulder and neck, drooping ears and horns that curve up and back. It is naturally grey but can be red in colour.

Simmental

The Simmental breed, yellow or red with white head and markings, takes its name from the Simme valley in Switzerland. It has become popular in Europe, Russia and especially Germany. The Simmental is a typical 'dual purpose' animal, producing excellent beef and milk.

Limousin

Originating from the area known as Limousin in the western central part of France, this breed is second to the Charolais in importance throughout Europe. It is smaller than the Charolais (though not as small as the Normandy, a most effective beef-producer) and long in the body. It has big muscles and provides very good meat. The largest of the important French breeds is the Maine-Anjou.

Charolais

The cream-coloured Charolais named after the Charolles area in Central France. It is an excellent 'triple purpose' animal because it produces good beef and milk, and is strong enough to pull a plough. Because it produces such good beef, it has been exported to the USA, to many countries of Central and South America and recently to Britain, where it has been cross-bred to improve other breeds. In the USA it is crossed with the Brahman, hence the Charbray breed which does well in hot, dry conditions.

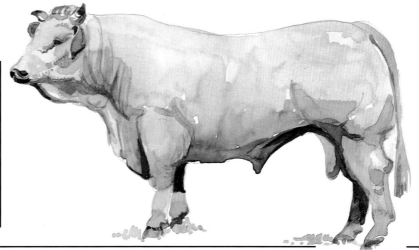

Beef Production

When selected dairy cows are mated with beef sires, the calves have 50 per cent beef and 50 per cent dairy characteristics, so producing more meat than a pure-bred dairy cow. Cross-bred calves grow fast because they inherit what is known as 'hybrid vigour'. The continental cattle have good muscular development of the back and hindquarters, which are the butcher's most valuable cuts. However, for sheer taste and quality, it is hard to beat our own Aberdeen Angus and grass-fed Herefords even if they do take a little longer to mature.

Worldwide consumption of beef exceeds 48 million tonnes a year.

DIFFERENCE BETWEEN A BEEF AND A DAIRY COW
The beef animal is muscular and well-covered with flesh. The dairy cow has a feminine head. She is long and carries little fat, which makes her look bony.

CATTLE MARKETS
Farmers have sold cows at market for hundreds of years so there are market towns all over the country. Before the days of cars markets were meeting places as well as the centres for trade and business for the farmers and their wives, who often lived in remote places and had few opportunities for social life. Today there are weekly livestock auction sales in most market towns. At some centres, like Banbury, Carlisle and Perth, There are breed society sales of pedigree stock in spring and autumn as well as the regular sales.

CATTLE SHOWS
There are five large cattle shows during the year; the Royal Show at Stoneleigh, Scotland's Royal Highland Show, the Royal Welsh Show at Builth Wells and the Balmoral Show in Ireland; in December the Royal Smithfield Show is held at Earl's Court in London.

There are many others. The exhibitors take enormous trouble preparing their animals for the show ring. They must be in top condition, well-fed, exercised and perfectly groomed. They have to be halter-broken and look their best when paraded in front of the judges.

BEEF CUTS
1 Shin
2 Brisket
3 Flank
4 Top Rump
5 Leg
6 Silverside
7 Rump
8 Sirloin
9 Ribs
10 Chuck and Blade
11 Neck
12 Oxtail

LEATHER

Leather is the treated skin of an animal. The treating process is called 'tanning'. Skins are bought in their raw state by tanners and are graded according to size - a hide is the skin of a large animal, such as a cow, bull or horse. The skin of a calf is softer than a hide; a skin comes from the smaller animals such as young calves, goats, lambs and sheep.

The tanners remove the hairs and clean them in various stages with chemicals. Hides that are thick are split into layers, rather like peeling a piece of paper off a note-pad, and the layers are called Outside, Top Grain, Deep Buff, Split, Slab and Insides. Skins are thinner and smaller and so provide fewer layers; those from small animals are not layered at all as they are very thin. Most skins are dyed but some, like pig skin, are left in their natural colour.

HAMBURGER

This is made with minced beef. It can be mixed with onion and herbs and bound together with raw beaten egg. The flat, round shapes are fried or grilled and usually eaten in a bun. The largest recorded hamburger was made in Cape Town, South Africa, in 1985. It weighed 2,270.66kg (5,005lb 13½oz), had a diameter of 7.10m (23ft 3.5in) and was grilled to provide 5,750 portions.

VEAL

Some calves are reared for veal; they are ready for slaughter at between 13 and 20 weeks old. Calves for veal have to be housed throughout their lives. The meat is very tender.

FANCY GOODS

Brief cases, purses, wallets and picture frames, are made from the Top Grain of cattle, horses and pigs.

UPHOLSTERY
The Top Grain from cattle, horses and pigs is used to cover chairs and stools and to upholster the leather seats of some deluxe cars, here a Rolls Royce.

BOOK BINDING

The book-binding trade uses goat skins, usually imported from Nigeria, India and Pakistan. Sometimes sheep's skin is used but is considered inferior in quality.

SUEDE, 'CHAMOIS' AND KID
Suede comes from the Deep Buff and Split layers and the suede effect is achieved by 'roughening' the leather on one side. Jeans, handbags and clothes are often made this way. Chamois is the under-Split of sheepskin; it also makes an excellent cloth for cleaning windows. Kid skin is made into clothes and of course kid gloves.

DO YOU KNOW?

The word 'beef' comes from the French word 'boeuf' meaning an ox.

Worldwide consumption of beef exceeds forty-eight million tonnes per year.

Until the 19th-century specialisation into beef or dairy farming, all cattle provided meat, dairy products, leather and pulling power. Until the last two hundred years, cattle were used as draught animals to pull ploughs and carts and were only slaughtered at the end of their working lives so most beef was of very poor quality and rather tough. In the Third World cattle are still used principally as draught animals.

SOAY SHEEP
This is a primitive breed, which came from the island of Soay in the St Kilda group, 50 miles off the west coast of Scotland. It is thought to be related to the wild Mouflon and looks rather like a goat. It does not need to be shorn; the wool falls off naturally like a blanket in the summer.

JACOB SHEEP AT CHATSWORTH
Jacob sheep are an ancient breed, mentioned in the Bible in Genesis, Chapter 30, when Jacob took all the spotted and black sheep from the flock. It is believe they came from the Holy Land, via the Mediterranean coast, to Spain and eventually to this country. They were kept for decoration in parks an became very rare. Now they have gained popularity and have commercial value. There is a flock of 5 breeding ewes and rams at Chatsworth. Jacob sheep are friendly. Their fleece is piebald (black and white) when they are born but the wool fades with time and they become brown and white. They have or 2 sets of horns. The ewes are excelle mothers and the lambs very strong. At Chatsworth lambing takes place before the main flock at the beginning of February, when they are brought into one of the farms. They are shorn in June and much of the wool is sold for handmade good The wool does not need dyeing as it has a wonderful natural colour, varying from white cream, grey, through to brown and black, and produces the famous Jacob tweed. In Septemb the surplus Jacob sheep are sold in the big shee sale at Chatsworth. New rams are bought when is necessary to change the blood.

Suffolk
One of the shortwool Downs breeds used primarily as breeding rams for crossing with ewes of other breeds and crosses to produce meat lambs. The Suffolk is a thick-set, early maturing sheep with the typical black face and legs of the Downs breeds.

Blue-faced Leicester
One of the Leicester breeds developed by the 18th century livestock improver Robert Bakewell. The Blue-faced Leicester is a longwool type which has influenced all the British longwool breeds. Blue-faced Leicester rams are crossed with Swaledale ewes to produce a mule, the most popular type of ewe used on the rich lowland farms to rear 'fat' lambs.

Clun Forest
Another of the upland breeds, and one of the few pure breeds which produce lambs of sufficient quality to rear on for meat. The ewes are prolific and fast-growing, and are frequently used for lambing on low ground farms.

Jacob
The famed sheep of biblical times, and named after the shepherd Jacob. Both sexes are horned - those of the male being particularly impressive.

Sheep

Because of the various types of terrain on which they must exist, British sheep have evolved into three distinct types, and over fifty different breeds.

The *longwools* are native to the fertile areas where food is plentiful, and the weather relatively mild. They are a large, hornless sheep with thick, heavy fleeces and white legs and faces.

The *shortwools*, most of which are also hornless, are a smaller, quicker-maturing animal with dark legs and faces, and have a finer, shorter, creamy fleece, sometimes mixed with brown or black. They produce excellent meat and thrive on dry, light soils.

Upland breeds must eke out a living on the sparse mountains and moors of Wales, Northern England and the Scottish Highlands, where the weather is foul, the soil poor and food scarce. They are small, active, very hardy and make excellent mothers.

Most of these breeds were refined during the Agricultural Revolution of the 18th and 19th centuries, when a sheep's fleece was of such value that a farmer could earn a comfortable living from producing nothing but wool. But today modern lowland sheep farming is geared almost entirely to the production of 'fat' lambs for the meat market.

The breeding ewes are often *mules*. A mule is a product of mating a Blue-faced Leicester ram with a Swaledale ewe. Another popular ewe is a *masham* (Wensleydale x Swaledale). Mules or mashams are then crossed with one of the Downs or Suffolk rams, to produce fast-maturing lambs.

The Chatsworth Flocks
Mules and mashams may be seen at Chatsworth, both in the park, and in the Farmyard. The Chatsworth flocks are shorn in early June, and income from the sale of the wool represents about 5% of the total income from sheep. The lambs born on the Chatsworth Estate are weaned in July and kept until September. Those not needed for flock replacement are sold at the annual sheep sale to buyers from all over the country.

FACT FILE: SHEEP

Male: Ram **Female**: Ewe **Young**: Lamb

The words used by flock masters and shepherds to describe the sheep of different ages and sexes are incomprehensible to most people, and no wonder!
  Here are some: hogg, hogget, pug, for a ram or wether (castrated lamb). In the north country the same animal can be a tup-lamb, tup hog, diamond tup, then a tup. Ewe lambs become hoggs, gimmers or chilvers. In Derbyshire yearling ewes are theaves or shearlings. Sometimes they are called after the number of teeth they have, which tells their age – two-toothed, four-toothed and full-mouthed at four years old.
  In old days the shepherd counted his sheep thus: 'Onetherum, twotherum, cocktherum, qutherum, setherum, wineberry, wigtail, tarrydiddle, den'.

Gestation period: 151 days

Welsh Mountain
The smallest of our indigenous breeds, the Welsh Mountain, of which the black is a variation, is a tough breed of the upland type. Known for its ability to survive and breed under extreme weather conditions and on the poorest of grazing, It tends to inhabit areas which other domestic animals find impossible. The ewes are often crossed with a Southdown ram to produce exceptionally good small lambs for eating. Because of their environment, Welsh Mountains rarely rear more than one lamb at a time.

Merino
Originally bred in Spain in the 15th century, Merinos are the pre-eminent producers of the finest quality wool. Not a good meat producer, and prone to foot problems in high rainfall areas, they thrive in the hot, dry climate of Australia, where Merinos are reared in vast numbers on the huge sheep stations.

Swaledale
Although bigger than the Welsh Mountain, the Swaledale is one of our smaller sheep breeds, and just as hardy as some of the other upland breeds. The Swaledale has a coarse fleece with a high content of kemp mixed in with it, and this low quality wool is used for the manufacture of carpets.

A Ewe's Calendar

The ewe's year starts in August and September when her health is checked and teeth, feet and udders are examined. Those not fit for breeding are sold – called 'cast' or 'culled' ewes. The breeding flock is moved on to good, fresh pasture for a few weeks to get them fit before mating (called 'flushing'). In the winter, when grass is in short supply, the in-lamb ewes are given supplementary food in the form of hay, silage or root crops and concentrates. Most ewes live outside all year round but it is becoming more popular with farmers to house the ewes for 2 or 3 months before lambing.

TUPPING

Farmers in the South of England prefer to mate their ewes with a ram in August and September to produce early lambs, which can be sold at higher prices. We cannot do this at Chatsworth because there is not enough grass for them to eat until mid-April. The rams are usually run with the ewes during October (known as tupping); the rams are fitted with a raddle (a harness with a pad across the front that is covered in a coloured waxy substance). When they mate with an ewe the raddle leaves a coloured mark on the ewe's rump so the farmer can tell which ewes have been mated and are likely to produce lambs.

LAMBING

Before lambing begins farmers prepare their yards, constructing pens with straw bales and hurdles. The lambing period is the busiest time for sheep farmers; they are often on duty all night ensuring that ewes and their lambs are all right. Most ewes lamb naturally and easily but the shepherd notices a ewe that is a long time in labour because the lamb is presented the wrong way round or if a leg or its head is twisted back and she needs help. The new born lambs are put in a pen with their mother to establish a family bond.

DIPPING

By law sheep must be dipped once a year to disinfect them against Sheep Scab, a dangerous infection, and to kill the parasites (lice and ticks that live in the fleece or just under the skin). This has to be done between 25th September and 3rd November. At Chatsworth the sheep are dipped twice a year; against blow fly in July when the fleece is short after shearing and then again after 25th September. Each sheep has to be completely submerged and remain in the solution for one minute.

FOSTERING

If triplets are born, one is often small and weakly. The mother may not have enough milk, or one may be rejected which the shepherd 'fosters' on to a ewe whose lamb has died. It has to be made to smell like her own lamb for her to take to it. When it replaces a lamb which has died, the skin of the dead lamb is fitted on to the foster lamb and so the ewe accepts it as her own. Some orphan lambs are bottle fed by the farmer.

SHEEP TEETH

Lambs are born with small milk teeth, which are pushed out when their permanent teeth come through. Sheep are nibblers; they only have teeth on their lower jaws and have a fairly hard gum on the top against which they chew. Telling the age of a sheep from their teeth is fairly accurate.

They get their first 2 permanent teeth aged around 15 months, when they are known as 'shearlings'.

The following year they have 4 teeth.

Each successive year they grow another 2 teeth until they have 8, and are known as 'full mouthed'.

Sheep's teeth continue to grow even when they have a full set. They either break or drop out as they get old. A 'broken mouthed' ewe is one that has lost teeth. Eating becomes difficult and the ewe may be culled from the flock at this point, when she will be aged between 7 and 10 years and have had several lambs.

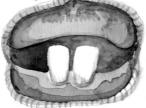
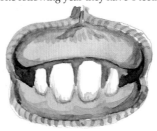
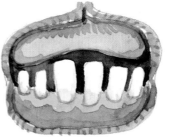
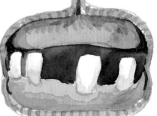

DOCKING AND CASTRATING

Lambs have their tails docked to help prevent blow flies from laying their eggs around the tail area. Also the ram lambs to be sold as meat are castrated. The shepherd puts tight elastic rings around the top of the lamb's tail as soon as it is born. This stops blood circulating in the tail and it drops off after 2 or 3 weeks. It does not hurt the lamb. The male lambs are castrated in the same way.

WEANING

At about 14 weeks, the ewe's milk begins to dry up and the lambs are weaned (taken away from their mothers). The farmers select the best ewe lambs to keep for breeding and the rest are fattened for sale or slaughter.

DISEASE

Ewes are injected with vaccine when they are pregnant to reduce the risk of the many diseases that can threaten their lambs. The antibodies in the vaccine continue to be transferred to the lamb after birth through the ewe's milk. All ewes and lambs are prone to having worms and must be dosed against them. Foot rot is commonly caused by wet ground. The shepherd has to trim their feet and put them through a medicated foot bath to guard against this.

SHEARING

Shearing takes place once a year during May and June, after weaning and when the weather is warmer. The ewes and rams are shorn, the lambs are not. Flocks kept indoors are often shorn in January to prevent them getting too hot. Shearing is done by experienced groups, often from Australia and New Zealand who tour the farms with their own equipment. Shearing takes place under cover and electric clippers are used but it does not hurt the sheep. About 200 sheep are shorn per day by each shearer, the expert can manage as many as 300, and the world record is held by a New Zealander, who in 1990 sheared 501 sheep in eight hours.

SHEARING IN THE OLD DAYS

The shepherding staff of a lowland farm pose with their newly-shorn sheep outside the shearing shed. In these days the sheep were shorn using hand-operated shears, a skillful and backbreaking task. Note the head shepherd's badge of rank – his bowler hat!

Wool

Wool used to be one of the main sources of wealth in this country and magnificent 15th and 16th century 'wool' churches in East Anglia were built as a result of the wool trade. The Cotswolds also had many 'wool' churches. Throughout the 17th and 18th centuries, wool continued as a valuable product in Britain; now the huge amount of quality wool produced by Australia and New Zealand together with the invention of man-made fibres, has made wool of secondary importance to meat in sheep farming. Recently there has been renewed interest in traditional uses of wool.

Quality in wool depends very much on climate - temperature and rainfall – from the soft, fine wool of the Shetland to the coarse fleeces of the Swaledale. The fibre is different from human hair in that it has a core covered with tiny scales and the fibre is wavy and not straight. When spun together, the scales and waves stick together producing skeins (or yarn) that are both strong and stretchy. The fineness of the fibres are assessed on the Bradford Count, that is to say long fibres with a lot of outer fleece are coarse and have a low count but the shorter fibres with more inner than outer fleece have a higher count, despite the short fibres.

SHETLAND SHEEP
A small hardy breed, the Shetland lives on the hills all the year round. It has very fine wool, which is used to make traditional Shetland-patterned jumpers. The fleeces may be white, brown or black.

THE FLEECE
The fleeces go to the Wool Marketing Board who sell them on to wool brokers. They are graded and sold to mills and private individuals. Every fleece must be washed to remove the dirt and grease. The rough edges of the fleece (known as the skirt) are removed and it is then processed into yarn; the longer wool being used to produce the smooth finish of worsted and the shorter wool providing the fuzzy texture of knitting wool. The fleeces weigh about 2.4 kg each and those of the rams are slightly heavier.

KNITTING AND WEAVING

Wool fabrics are made from yarn that has been knitted or woven. Knitted fabric is made by looping the yarn continuously either by hand with needles or by machines. Hand-knitting has been practised since the 5th century BC and examples have been found in Egyptian tombs dating from that time. Weaving is when two sets of yarns (one vertical called the warp and the other horizontal called the weft) are interlaced together at right angles; it was developed during the Industrial Revolution in the latter part of the 18th century. Different machines produce the various types of woven fabrics, which include:

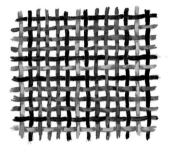

Woven yarn which provides us with clothing and upholstery fabrics;

SHEEP IN JACKETS

Occasionally, on smallholdings, when a Shetland sheep has lambed and is in poor condition, she runs the risk of losing her wool and then her milk supply. Old jumpers are put on to the ewe to prevent this happening and to keep her warm.

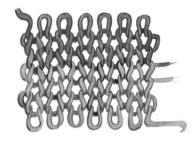

Knitted fabrics such as jumpers, skirts, hats and gloves;

Carpets which can be tufted (shown) or woven.

COLOUR AND DYEING

Dyeing can take place before or after spinning. Colours obtained from plants have been used in 'natural' dyeing for thousands of years; chemical dyes (anile) were introduced in the 1850s and wool mills can dye huge amounts at one time (sometimes as much as 1 tonne in fleece-form). There is a swing back to natural dyes and their softer colours which blend so well together.

THE WOOLSACK

The Lord Chancellor in the House of Lords sits on a large, square bag of wool, covered in red cloth called the Woolsack. The seat has no armrests but does have a backrest. The Woolsack was placed in the House of Lords during Edward III's reign in the 14th century to mark the importance of the wool trade in England at that time.

Mutton and Lamb

Lamb is a very popular meat in this country. Lambs are killed at about 5 months old when they weigh about 37 kg, according to the breed and the way they have been fed.

When the butcher has cut off the unsaleable parts from the carcase he is left with about 17 kg of meat. It is difficult to find mutton (meat from a sheep 12 months old or over) in butcher's shops now, partly because housewives prefer the younger lamb and smaller joints, and partly because the farmer needs to sell most of his lambs in the autumn so he does not have to keep them through the winter.

If it is well-cooked, mutton is delicious, from the classic leg of mutton with onion sauce to stews, hot-pots and curries which have much more taste than those made with lamb. Sheep meat absorbs flavours well and herbs such as rosemary and spices suit it well.

STONE TENTS
On the Chatsworth estate some of the small barns are used for camping by the boy scouts and girl guides; they were once used by farmers for wintering young stock.

Sheep's Milk
Ewes may be kept for their milk which, like goat's milk, is an excellent substitute for cow's milk for people who have allergies. Ewe's milk is richer than cow's milk. Dairy sheep must as be well fed as a dairy cow. They are machine-milked twice a day. The main dairy breeds are the Friesland (from Holland) and British Milk Sheep. Frieslands can give up to 70 litres of milk during a lactation (milk producing period) which lasts for 8 – 10 months. Just over 2 litres of milk makes 0.5 kg cheese.

Lamb Cuts

1 Head
2 Scrag
3 Shoulder
4 Neck
5 best end
6 loin
7 middle neck
8 Saddle
9 Leg
10 Breast
11 Chump Chop
12 Cutlet

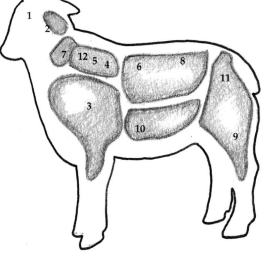

SHEEP'S CHEESE

Shepherds and monks have made cheese from ewe's milk for hundreds of years and the practice is common in Europe, the most famous is the French Roquefort, matured in caves. Sheep's cheese has only recently become popular again in Britain. Both hard and soft cheeses are made (*see page 15*).

Sheep's milk ice cream
Ice cream can be made with sheep's milk in the same way as cow's milk.

THE SHEPHERD AND HIS DOG

The collie was developed as the best dog to work with sheep and cattle during the 18th century. The rough-coated collie was originally used to guardand herd the sheep and the smooth-coated type was used to drive stock to market. The most outstanding as a sheepdog is the long-haired, black and white Border Collie that has been used by shepherds in the borders of England and Scotland for over 300 years. The collie is a very active dog with a tapered head, small erect ears and almond-shaped eyes. It is famous for its loyalty to its handler.

Haggis
Haggis is the heart, lungs and liver of a sheep (or a deer), minced together and mixed with suet, oatmeal, onions, salt and pepper and then boiled in a skin like that of a sausage . It was a popular dish all over Britain until the 18th century but is now considered purely Scottish, and is eaten on Burns night (Robert Burns, the Scottish poet, born on 25th January 1759).

SHEEPDOG TRIALS

Sheepdog trials test the dogs and their efficiency in moving and controlling the sheep. They work singly or in pairs, fetching and driving sheep over a marked-out course and herding them into a pen. Trials became popular in the 19th century; since the television series of '*One Man and His Dog*' millions of people have been able to see the way they work the sheep. Farmers and shepherds compete with their dogs, and trophies are to be won in many local, national and international trials.
The dog is commanded by his handler. He sets off in a wide run so he can move in behind the sheep without disturbing them too much. He drives them steadily towards the handler who continues the command with a whistle or his voice. The course usually consists of a few gates and pairs of hurdles through which the sheep are driven and the dog has to separate (shed) the sheep, sometimes two at a time, while controlling the rest of the flock and eventually pen them in a small enclosure. Obedience is paramount and the tests are not easy, for sheep are naturally nervous and can be obstinate.

Pigs

Pigs were domesticated in China as long ago as 1100 BC and farmed in other countries soon after. The Ancient Greeks and Romans ate pork and it became popular in England during the Middle Ages when huge tracts of land were covered with forests. In the 18th century, around the time of the Industrial Revolution, an English farmer, Robert Bakewell, crossed imported Chinese breeds with a wild pig from Europe to produce a 'white' breed that is so popular today.

Pigs are intelligent creatures, and, despite their reputation, they are naturally clean. They mess away from their sleeping area and the pens are designed to allow for this. They are unable to sweat, so they need a lot of water to prevent them becoming too hot; they like to wallow in mud to keep cool. They do not chew the cud, like cows, sheep and goats, and so are banned as food by practicing Jews, whose religion only allows them to eat meat from animals which 'divide the hoof and chew the cud'.

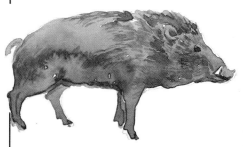

Wild Boar

The wild boar, with its bristly coat, brown or black and grizzly (with grey hairs) in colour, has always been considered an excellent animal to hunt because it is fierce, fast and strong. King Richard III (1483-85) had the boar as his emblem for his badge. The wild boar is still hunted in parts of Europe and India.

Tamworth

Tamworths have long snouts and look more like wild boar than any other pig. The Tamworth originated in Staffordshire, but is now farmed in the USA, Canada, Australia and New Zealand as well as England. Its colour varies, but it is usually golden-red. It is a large animal, producing good lean meat and often cross-bred for its bacon.

Gloucester Old Spot

This is one of the old-fashioned, hardy breeds which used to be reared and kept in sties by country people. We keep Gloucester Old Spots in traditional, non-intensive conditions at Chatsworth.

Berkshire

This pig was originally established in 1770 in the English county after which it is named. It is mainly black, with white on its face, legs, and the tip of its tail, and has erect ears that point forward. The Berkshire is bred for pork in England and Japan, but in Australia and New Zealand a larger variety has been developed for its bacon. Pig Wig in Beatrix Potter's *Pigling Bland* was a Berkshire.

Landrace

Pigs used to be reared for their fat (lard) as well as for their lean meat, but with the increasing use of vegetable fats such as margarine, the meat became more important. In the 1930s Swedish farmers crossed the newly developed Landrace with other breeds to produce bigger pigs which gave less fat and more lean meat, and had larger litters. Now almost all pig-farming countries have the Swedish Landrace. It is a white pig with a long back and lop ears.

FACT FILE: PIG

Male: *boar* – normally kept for breeding

Female: *gilt* – young female before her first litter
sow – full-grown, used for breeding

Male or female: *porker* – pig from 50 to 67 kg
cutter – pig which weighs 68 to 82 kg
baconer – pig weighing 83 to 101 kg
heavy pig – a pig weighing over 101 kg when it is slaughtered

Young: *piglet* – normally between 6 and 15 are born in a litter
runt: the smallest in the litter
weaner – piglet weaned aged 6-8 weeks, and sold for fattening; they weigh around 25 kg

Gestation period: 112 days

British Saddleback

This is a comparatively new breed, the result of crossing two very old breeds – the Wessex Saddleback and the Essex Saddleback. It takes its name from the white strip across its shoulder which resembles a saddle. It is black apart from this strip, lop-eared - that is, its ears flop forward - and it can vary in size. The pure-bred saddleback has a lot of fat, so it tends to be crossed with white breeds to improve the quality of its meat. The sows are hardy and do well under non-intensive breeding conditions out of doors. They are also good mothers. For these reasons they are increasing in numbers.

Duroc

The modern Duroc pig is the result of a cross between the Jersey Red and the Duroc of New York. It is said that Columbus introduced the 'red' pig to America on his second voyage. The Duroc can be various shades of red, but has no black. It became very popular in the USA in the 1930s, and has remained so ever since.

Pig Keeping and Breeding

Sows are mated for the first time when they are about 6 months old. There will be 10 - 12 piglets in the average litter, and a sow can produce about 5 litters every 2 years. The largest recorded litter in England was born in 1979 and contained 30 piglets.

The boar first mates when he is about 7 months old. He can 'serve' up to 40 sows in a year. Sometimes farmers prefer to use artificial insemination instead of a boar. Artificial insemination (A.I.) means that the egg is fertilised by semen reaching it through an artificial agency. With pigs this is achieved not by the usual trans-ference of the sperm from the male to the female during mating, but by inserting a small tube of semen into the sow's uterus.

COTTAGE STIES

The stone-built pig sty at Chatsworth Farmyard is a good example of how pigs were housed when most country people kept a pig to provide bacon and hams for the family. They were fed on meal, potatoes and kitchen scraps. In the days before refrigeration, pig meat was cured with salt and smoked to prevent it from going off, so providing the family with meat all year round. Many old farmhouses and cottages have big hooks in the beams of the kitchen where the sides of bacon and the hams were hung.

NATURAL FARROWING

As soon as a sow farrows (gives birth), the piglets find her teats by instinct and begin to suckle. It is a case of survival of the fittest, as the bigger piglets push the smaller ones aside, so it is important to make sure the little ones manage to feed too.

Large White

There are more Large Whites in the world than any other breed of pig. It is hardy and thrives in all climates. It converts the food it eats to the highest daily gain of any breed. They produce best quality meat for the pork, cutter, bacon or heavy pig market. The boars are used for crossing with other breeds. The sows are docile and good mothers.

FARROWING

Pigs may be farrowed in crates on intensive pig-breeding farms. The crate prevents the piglets being accidentally killed by the sow lying on them.

EXTENSIVE REARING

The hardier breeds do well under this system, where they are kept out of doors and have relatively cheap corrugated iron shelters known as arks. The pigs are allowed to forage - dig with their snouts - for some of their food. In pig rearing, food accounts for 75% of production costs, so the savings on food and on the expense of intensive housing , together with the fact that it is a healthier way to raise piglets, make the extensive system very attractive. The problem is that the vast majority of pigs bred today are reared for meat, and they do not reach killing weight so fast under the extensive system because they grow more quickly when in a warm house. However the outdoor system is gaining fast in popularity as people are gradually returning to more natural methods.

TRUFFLING

ruffles are a fungus elated to mushrooms. They are considered a reat delicacy and fetch normous prices. They row among the roots f oak trees and beech ees in France and ngland, but because ey tend to lie about) cm down into the round, they are ifficult to find. Pigs ave a keen sense of nell and can be ained to dig up uffles, and so can ome breeds of dog. he Périgord region of rance is famous for its igs and truffles.

DEVELOPMENT OF THE PIG

piglet weighs about 1.3 kg at birth and at hatsworth feeds from its mother for a eek or so before it is weaned on to oncentrates, which are specially made olid foods.

At the age of about 8 – 10 weeks, pigs which are not being used for breeding may be sold as 'weaners' to farmers who specialise in fattening them for market. They weigh about 25 kg at 10 weeks.

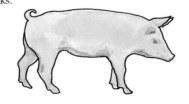

young boar that is to be used for reeding is also normally sold at about 7 onths. It is important that these boars are e offspring of an excellent sow and that ey are strong and fit, as breeding boars e expected to sire large numbers of iglets.

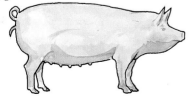

Pigs being fattened for the butcher are slaughtered when they weigh 40 - 67 kg if they are to go for pork to your local butcher; at 70 - 80 kg (when they are known as cutters) for the supermarket; or at 83 - 101 kg to be sold for bacon. They are then 6 months old.

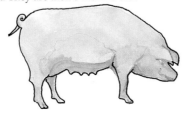

INTENSIVE REARING

Most pigs are reared intensively in specially-built pig houses where the temperature is thermo-statically controlled to encourage rapid growth. This ensures that most of their food goes to increasing weight rather than keeping warm. The food is provided automatically from 'hoppers'. Under this system neither the sows nor the growing pigs go outside. They are easy to keep clean as they dung in special areas which are sluiced down. Sometimes they are kept in strawed yards and not in pens.

Pork and Bacon

Pigs bought at market are nearly all pork pigs or cutters. However, most pigs nowadays are bought direct from farmers and go straight to the bacon factory or slaughterhouse.

There are 2 traditional methods of treatment for bacon. One is known as the Wiltshire cure: the pig is cut in half lengthwise and 'cured' by soaking it in heavily salted water known as brine. The result of this is called 'green' bacon. Green bacon is then wood-smoked to obtain smoked bacon. The other is dry-curing: a slower process where rough salt and a little saltpetre is rubbed into both sides of the pig.

PRODUCTS OF THE PIG
Apart from the squeal, every single part of the pig is used.

BACON CUTS
Due to supermarket demand, bacon nowadays is usually sold ready-packed. Normally the pig's carcass is divided up into cuts in the factory and the fat trimmed.

Bacon
Bacon comes from the back and belly of the pig. Like ham, it is salted and may also be smoked.

Ham
This is the best part of the hind leg , salted or dried.

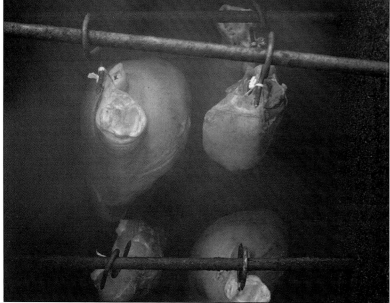

SMOKING HAMS
Few people smoke hams commercially. In a deep metal box, often surrounded by bricks, sawdust made from mixture of woods (including hickory and oak) is burned and the smoke puffed around the hams. Ham and sides of bacon take 24 hours to smoke.

PORK CUTS
Of the various pork cuts, the leg is usually roasted; the loin is fried, grilled or roasted; the tenderloin or fillet is fried, grilled or baked; the belly roasted or boiled; belly slices are grilled and the bones are removed to make barbeque spare ribs; the hand roasted; the neck roasted as spare ribs. The blade and or shoulder bone may also be roasted; the head is used for brawn or sausages; the tail salted; and the trotters salted or boiled.

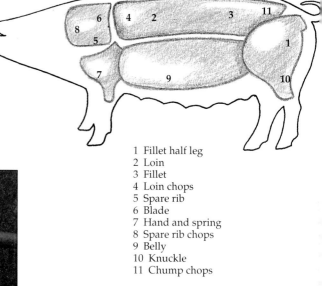

1 Fillet half leg
2 Loin
3 Fillet
4 Loin chops
5 Spare rib
6 Blade
7 Hand and spring
8 Spare rib chops
9 Belly
10 Knuckle
11 Chump chops

Chop
A chop literally means a piece of meat that has been cut off. is also known as a cutlet. The loin chop also contains the tenderloin or fillet (like a T-bone steak).

SHISH KEBAB
This is a Turkish dish now familiar in Britain. 'Shish' is the Arabic word for skewer. In days past, the sword was used to skewer the meat, which was then held over the fire to cook.

Lard
Pigs' lard (fat) is much used in cooking, especially for roasting meat.

...air

...gs' hair is used to provide the bristles for paint brushes.

...lands

...gs' glands produce important drugs for medical purposes ...d pigs' heart valves are used in surgery.

...AUSAGES

...usages are made of finely chopped pork (or beef or a ...mbination of the two) together with rusk ...readcrumbs), herbs and salt. The mixture is pushed ...to a skin. The meat can be fresh, smoked or cured.

Pigskin
Pigskin is a luxurious form of leather used to make wallets, purses, umbrella handles, small attaché cases, photograph frames and writing cases.

BONEMEAL
Cattle and pigs' bones are ground up to make bonemeal, used as a fertiliser on garden plants, especially roses, and it is also used as a dietary supplement for in-foal mares and young stock.

...n early illustration showing the butchering of ...e family pig.

Roman mosaic showing
a cockerel

Welsummer

This light breed comes from Holland and is no found all over Europe. T hens seldom become broody and are good lay Their eggs are dark brow but not as dark as the Maran's eggs. Light breeds considered to be flighty birds that flap around when handle but like the White Leghorn, the Welsummer bird becomes tam with handling.

Chickens

Chickens are descended from the wild Asian jungle fowl. The cock birds of some strains, especially Old English Game Fowl, are aggressive as their instinct is to kill rival cocks and protect the hens. Cock fighting is still a popular 'sport' in many countries, but is against the law in Britain.

In the last 150 years chickens have been selectively bred to produce eggs *or* meat, but there are some dual-purpose breeds. Naturally a hen, like all other birds, lays a clutch of eggs in the spring and then 'goes broody', which means she sits on her eggs to keep them warm and allow the embryo chick inside to grow. She leaves her nest once a day for food and water, returning to sit on the eggs before they go cold. The chicks hatch out of their eggs by pecking through the shell with their beaks. They start to peck at their food at a few hours old. They must be kept warm either by the broody hen or under an electric lamp until their adult feathers have grown, at about 6 to 8 weeks old.

The best layers are hybrids of the light breeds (light in weight, not necessarily in colour). They convert most of the food they eat into eggs, not body-weight. The layers are kept usually for only one year, as they lay fewer eggs in their second year but eat more food, and are not so profitable. We buy a new lot of hybrid layers from a specialist breeder in the spring when they are about 20 weeks old and are at 'point of lay'. Their eggs are sold in the Farm Shop.

CHATSWORTH FARMYARD
In the Farmyard at Chatsworth we keep some rare breeds of poultry like Buff Cochins which have feathered legs like trousers, Partridge Laced Bantams, Appenzellers and crosses of these breeds Silkies are among the most attractive. They have fluffy feathers, even on their legs, and blue/black combs. When they go broody they are excellent sitters and make perfect mothers and so they are often kept to hatch eggs from other breeds. Every year one or two of our hen lays a clutch of eggs in the bushes where she can't be seen and produces up to a dozen chicks which she brings proudly home to the yard, where you will see them, as they are free to go where they like.

The chickens and ducks have to be shu in houses at night because of the danger of being killed by foxes.

Rhode Island Red

Originating in North America, the Rhode Island Red must be the most famous breed of chicken. It provides excellent meat, and is a good layer (a successful dual-purpose bird). The eggs are light brown and it is claimed that it can lay up to 300 eggs per year. The cockerels are well-known for their aggressive nature.

Light Sussex

The Sussex chicken is a popular English breed, the best known being the Light Sussex which is considered an excellent table bird and egg-layer, and a good mother.

Old English Game Fowl

An upstanding, broad breasted breed originally kept for cock fighting and often seen in old pictures.

FACT FILE: CHICKEN

Male: *cock, cockerel* – a bird kept for breeding
Female: *hen* – adult, egg-laying bird
broody hen – female sitting on eggs to hatch them
Young: *chicks*

pullet – young hen from the time she begins to lay until the first moult

Hatching time: 21 days

Maran

The Maran bird comes from France and is a heavy breed, the most common being the black and white Cuckoo Maran. Its dark brown eggs are very popular but the shells are hard and can cause problems during hatching (when the chick chips its way out of the shell). They make good mothers and provide good meat too.

Silkie

The Silkie chicken is often called a Bantam (*see page 37*), but is, in fact, a large bird. It originated in Asia and is classified as Light. The white breed is best known.

DEEP LITTER

The idea of deep litter as a covering for the floor of a chicken house came from the USA in 1949. About 8 cm of straw, sawdust or peat moss was put down over a thin layer of horse manure. When mixed with chicken droppings, this produced a dry material that could remain in place for up to 2 years, provided there was suitable ventilation to stop the strong smell from harming the birds. Variations of this method are used now, as it is an easy and economical way of keeping a chicken house clean.

PREENING

All birds preen – groom – themselves to keep their feathers in good condition. They do this by rubbing their beaks along the oil glands at the base of their tail feathers, and then spreading the oil over the rest of their plumage.

More on Chickens

Although chickens had certainly been domesticated 4,000 years ago, it is thought that they must have been hunted for their meat and eggs at least 1,000 years before that. Today, about 450 million chickens are sold in this country every year for roasting or boiling. Meat-producing birds (broilers) are bred in the same way as the egg-producers (*see page 38-39*). Chicken meat is sold fresh, chilled or frozen, whole or in portions and, although much of it is now pre-processed, a whole chicken remains a favourite for the Sunday roast. Chicken meat is valuable for those needing to diet, as it is high in Vitamin B, protein and minerals but low in the all-important calories.

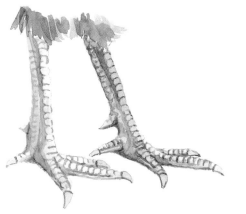

CHICKEN FEET

The colour of chicken's legs varies according to the breed.

The adult chicken has a claw on the back of its leg, called a spur.

They have three toes to the front of each foot, with longish claws. You will often see them scratching around a farmyard, looking for seeds and insects.

PECKING ORDER
There is always a boss in the flock, hence the expression 'pecking order', as the boss grabs the best of the food. This can lead to them pecking out each other's feathers. When this happens, especially in the case of birds known to be particularly aggressive, their beaks may be trimmed and the upper beak may even be removed. Bird's beaks are as strong and necessary as teeth are in animals.

COCK'S COMB
Like many of the males in the bird world, the cock is the more colourful of the sexes. As well as his bright plumage, he has a distinctive red crest on his head, a red face and two pieces of red skin on either side of his beak.

GUINEA FOWL

The Guinea Fowl, slate grey in colour and speckled with white spots, is reared all over the world, especially in Italy. It has short, round-shaped wings and flies well but prefers to run when frightened. The hen lays a clutch of eggs during the spring.

CHICKEN DISHES

Due to the increasing demand for fast food, a certain amount of chicken meat is processed and made into pies, burgers, casseroles and various other dishes. They are sold either fresh or frozen as 'heat and serve' meals. This is a Peanut Chicken Pastie.

CHICKEN MEAT

Birds bred for cock fighting have helped the development of chickens kept for meat, as the fighting birds needed to be strong and have broad breasts.

BANTAM

The bantam is a small variety of the larger bird. It is said to be named after a village in Java, an island of Indonesia in South-East Asia. Because of its size, it was easy to smuggle aboard the old trading ships. Bantams need comparatively little space, housing and food and, of course their eggs are small.

FERTILISED EGGS
On the yolk of the egg, surrounded by the white, is a tiny spot that will grow into a chick.
The egg is kept warm (either by the hen or in an incubator) and red lines appear in the yolk. The chick is starting to grow.

The chick's eyes can been seen as black spots. It feeds off the egg yolk and breathes through the porous shell.

At 9 days, the red lines that feed the chick with food and air are much bigger.

At 18 days it is possible to recognise the chick, packing into the sac of fine skin that protects it within the shell.

At about 21 days, the chick starts tapping at the egg shell and makes a hole. It breaks the shell open and comes out, covered in yellow fluff, very wet and tired. Soon the chick dries and starts running about, keeping close to its mother and learning from her.

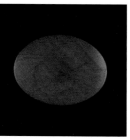 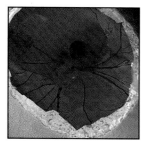 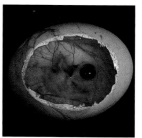 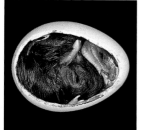 

NEW LAID EGGS
The yellow part of the egg in the middle is the yolk.

The white part surrounding the yolk is the albumen, but is usually called 'the white'.

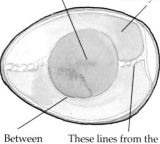

At the broader end of the egg there is a pocket of air called the air sac. This sac is divided from the rest of the egg by a thin membrance that has a texture rather like a balloon.

Between the yolk and the white of the egg is a layer of water.

These lines from the yolk to the wide and the pointed parts of the egg are called the chalazae and are twisted strings of membrane that keep the yolk in the right position.

Eggs

Spring is the natural time for hens to lay eggs and those of egg producing strains usually lay one egg a day throughout the year (*see Battery Hens*), averaging about 250 eggs per year. Normally the domestic hen will continue to lay one egg a day for 7 – 10 months. The most popular pure breed of egg-producing hen is the White Leghorn, which also holds the record of laying 371 eggs in 365 days during 1979.

The hen has to mate with a cockerel in order to lay fertilised eggs which will produce chicks. Fertilised eggs on commercial farms are hatched in incubators. They are collected twice a day and should not be washed before being put into the incubator as the egg shell is porous - that is it lets in water, air and light. The air sac is at the broader end of the egg, and is vital to the unhatched chick.

Egg Shells
All humans, animals and birds start life in an egg. Birds' eggs have hard shells to protect the babies because they are hatched outside their mothers. The shells of hens' eggs are less than 1 mm thick, and fragile. However, try to crack a fresh egg in your hand, squeezing it end to end, and you will find it impossible. This is due to its clever oval design which enables the shell to withstand the pressure. The furthest an egg has been dropped without breaking is 198 metres from a helicopter over a golf course, in Japan in 1979.

Colour of Eggs
The colour of hens' eggs depends on the different breeds: the Maran lays dark brown eggs, the Rhode Island Red brown eggs, the Light Sussex slightly tinted eggs and the American White Leghorn white eggs. The Araucana is an unusual breed and its eggs are blue. The colour of the yolk depends on the food the hens eat; a deep yellow yolk shows that the diet contained lots of protein but a pale yolk usually means that the food was not right.

CHICKENS IN THE FARMYARD

The chickens and ducks have to be shut in houses at night because of the danger from foxes. A farmyard would not be complete without at least a few chickens wandering around freely, even with today's intensive farming.

BATTERY HENS

Battery hens are kept in cages in a house where the temperature, ventilation (air supply) and light are regulated to make them feel warm and comfortable. The lights are kept on to trick the hens into believing it is springtime which encourages them to lay more eggs. Their droppings fall through wire and are carried away on a belt (so they are kept clean and less prone to disease); the eggs roll down this wire on to a tray and are carried away; and the hens' food is delivered on a belt that passes in front of them. Water is always available.

Public opinion is swinging away from keeping hens in such an unnatural way that they cannot perch or scratch or preen their feathers because the cages are so small. The eggs are cheaper and the customers have to choose between the taste of the free-range and the battery eggs.

CHICKEN FOOD

Chickens are often fed 'layers mash', bought already mixed, and grain. Free-range hens love greens such as cabbage leaves and kitchen scraps, and they scratch the earth for insects and grubs. A laying hen needs about 100g of food a day to feed herself and produce a 50g egg.

EGG PHRASES AND FABLES

Lots of expressions in the English language feature eggs:

To lay a golden egg
Success

Don't put all your eggs in one basket
Don't put all you have into one gamble

Teaching your grandmother to suck eggs
Impertinently advising those older and more experienced

Egging on
exciting someone to do something

Egg-head
An intellectual person (from the belief that bald people are clever)

As sure as eggs is eggs
Certain

Bad egg
Not a good person

Scrambled egg
The nickname for the gold oak leaves on the brim of a service officer's hat

Easter Egg
From ages past the egg has been a sign of fertility and has been a Christian symbol, representing Christ's resurrection. Hence eggs play an important part in Easter celebrations.

Egg and Spoon Racing
An American won the longest fresh egg and dessertspoon race in 1979 when he ran 45.86 km in 4 hours 34 minutes.

GRIT AND WATER
Hens need to eat grit, which is tiny bits of sand or stone, and crushed egg shells, partly to help form the shells of the eggs they lay and partly as an aid to digestion. Water must be available at all times.

Egg Shelling
In 1971, in Kent, two men, both of whom were blind, shelled 1050 dozen hard-boiled eggs in 7½ hours, and hold the record.

Ducks

Domesticated ducks are descended from the wild Mallard, which originated in China, with the exception of the Muscovy. They are large-billed waterfowl, generally divided into 3 groups - those that dabble, those that dive and those that perch. They walk with a waddle because their legs are set a long way back on their bodies to help them swim. Ducks were first kept in captivity in Britain in the 13th century, and the county of Norfolk has become famous for its duck rearing.

Ducks do not thrive under intensive conditions, so ducks reared commercially are kept in houses with plenty of space, or in large, open pens. These ducks will be killed when they are 7 to 9 weeks old, but many farms still have ducks waddling around free.

Mallard

The Mallard is a typical dabbling duck. Mallard make beautiful nests lined with soft 'down', their very fine under feathers. The nests are usually on the ground, hidden in rough places under bushes, but sometimes they nest in hollows or the forks of trees. When this happens the ducklings flop down to earth; they are so light they don't hurt themselves. The mother duck leads them to water where they are safe from vermin, but like all wild creatures they still have their enemies. Pike will eat them, as will herons. Because of so many dangers it is unusual for a Mallard to bring up all her brood. They are a famous game bird and are very good to eat.

DABBLING

Certain ducks dabble in mud or shallow water. They up-end themselves, and graze along the pond bottom for food. Beatrix Potter painted lovely pictures of white Aylesbury ducks 'up-ending' in *The Tale of Tom Kitten*.

DUCKS EGGS

Ducks eggs are larger and richer than hens' eggs and must be eaten very fresh. Because of their richness they are excellent for cooking. Ducks are not particular as to where they lay, and drop their eggs anywhere in their house or in their run. The egg shells are porous, that is they let in air, water and light, so should not be left in damp, dirty places. The Khaki Campbell is a wonderful egg-producer and can lay over 300 eggs a year.

Aylesbury

This is is the commonest breed kept in Britain. It is a large white duck with a yellow beak which has proved very succesful when bred commercially for its meat and its eggs. Aylesbury duckling is a delicacy found on the menu in most restaurants. They are quick growers, and are ready to kill at 7 to 9 weeks old.

Indian Runner

The Indian Runner was imported from India by a sea captain in about 1850. Its appearance very much resembles a penguin because the legs are positioned towards the back of its body, unlike other ducks, obliging it to stand very upright. Despite its small body it is considered a good table bird, the meat being similar to that of a wild duck. It is an excellent egg layer, producing at least 200 eggs a year that are white and the size of a hen's. It is a breed much admired at shows. The Indian Runner varies in colour but is mainly black, chocolate brown, fawn, white and fawn or white.

NESTING

Ducks kept where they wander free, whether on farms or in gardens, can lay their eggs literally anywhere. However, they do prefer to make a nest, usually on the ground under bushes.

During the incubation period (while she is waiting for her eggs to hatch), the mother turns them every so often. For the purposes of commercial breeding, the eggs are taken away from the mother, hatched in incubators and the ducklings kept warm under lamps.

WEBBED FEET

Ducks have webbed feet because they are water fowl and are agile swimmers but ungainly walkers. The webs, or skin between their toes help them to swim. Ducks that roost in trees have longer claws than those that swim a lot.

FACT FILE: DUCK

Male: drake

Female: duck

Young: duckling

Hatching period: 28 days except the Muscovy (35 days)

DUCKLINGS

Adults mate with a different partner each year. Ducklings can swim as soon as they are dry after hatching, but stay firmly with their mother, usually following her in single file, whether swimming or waddling along on dry land. Their feathers are fine and soft, like their parents' underplumage.

Domestic ducks are not very good mothers so their eggs are usually hatched in incubators or under broody hens.

Muscovy

Described by a Frenchman in 1670 as 'a wild Brazilian duck wth the bigness of a goose', the Muscovy was first found in South America. It builds its nest above the ground and its long claws help it to perch in trees, but it is not a good flier. The colours are all varieties of black, grey and white and it has a knob of red on its bill. They are greedy and bossy and so are best kept on their own as they will take all the food from smaller breeds. They make good eating when young.

GRAZING GEESE

Geese are grazers, they nibble grass even closer to the ground than a sheep does. They must have plenty of space to find enough fresh grass and can thrive on rougher herbage than other grazing creatures. They do not need much extra food in spring and summer but, when the grass is finished in autumn and when being fattened for the table, they are given extra food; usually grain with cabbage or kale added. They love apples and are often turned into orchards where they keep the grass down and eat the fallen fruit.

A good water supply is very important. They enjoy a stream or a pond, but can do without these as long as they have a trough deep enough to put their heads and long necks right under the clean water. They must have a dry shed to go into at night.

GEESE AS 'WATCHDOGS'

Geese only lay eggs in the spring, the natural breeding time. In this season the ganders can be very fierce, as their instinct is to protect the goose. They attack marauders with outstretched necks and make a loud hissing noise.

The ganders make excellent guards as they are very wary of strangers and cackle, hiss and flap their wings when approached. There is a famous story that they saved the Capitoline Hill in Rome from a surprise attack by the Gauls in 390 BC by cackling to wake the sleeping guards.

GOSLINGS

Goslings, like their parents, are good foragers and should be allowed to graze from the age of about 2 weeks, but they must have extra food in the form of barley meal, ground oats, bran, bone meal and linseed meal mixed with water. It is also important to protect the young geese from foxes, rats and crows during their early weeks and their enclosed runs should have wire netting on top and underneath.

It takes time for the gosling to mature and a goose should not breed until it is 2 years old; a gander can be as much as 4 years old before it is mature. When the gander is put with a number of geese (usually between 2 and 5) to mate, this is called a 'set'. Geese can breed succesfully for as many as ten years. The oldest recorded gander, George, lived in Lancashire and died aged 49 years 8 months.

Embden
The Embden - white with an orange beak, legs and feet - is perhaps the most popular breed in Britain. It comes originally from Germany and is prized for the quality of its meat.

42

EGGS

A goose's egg is larger than a hen's egg and some people consider it a delicacy. Goose eggs are not often sold commercially, and will be difficult to find unless you know someone who keeps geese.
Geese take a long time over their courtship.

```
FACT FILE: GOOSE

Male: gander

Female: goose

Young: gosling

Hatching time: 25-32 days
```

GEESE AT CHATSWORTH
Embden geese wander around the Farmyard. We let them into the paddocks and also have a pond where they can swim. There is a wooden hut by the pond where they shelter. Geese are very fierce (*see Geese as 'Watchdogs', left*), so, unlike ducks, they do not need to be protected from foxes.

THE FEAST OF MICHAELMAS
The Feast of Michaelmas, 29th September, was the traditional time for eating geese, when they were fat from the best of the summer grass. Goose Fairs were held in many towns. Now geese are becoming popular again at Christmas (as they were a hundred years ago) as an alternative to turkey. They are bred in their thousands on the continent of Europe where you see huge grazing flocks.

Toulouse
Like the Embden, the grey Toulouse goose is bred throughout the world for its meat. Males and females have similar plumage.

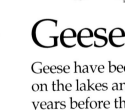

GOOSE FEATHERS
Before firearms were invented, goose feathers were used to flight arrows from the longbow. The 'down', or soft under feathers, are used to stuff the lightest and most luxurious pillows and duvets.
   Quill pens, the only way of writing with ink until metal nibs were invented, are made from the biggest feathers. The tube of the feather, which is where it is attached to the bird, is sharpened with a knife and the ink is held in the hollow.

Geese

Geese have been domesticated since 4,000 BC but were hunted on the lakes and swamps of Europe and North Africa for many years before that. The Romans bred white geese and took them with them to central and eastern Europe, where they are still found.

   Goose feathers were important in England during the Middle Ages: arrows flighted with them helped defeat the French at the Battle of Agincourt. By the 18th century, geese were being bred for their meat, but this is now a minor farming enterprise in Britain: it is only in Germany, Austria, Poland and parts of France that geese are reared commercially on a large scale.

Turkeys

The turkey is the largest of the game birds and is native to North and Central America. It was first domesticated by the Aztec civilisations and was brought to Britain in the 16th century. An old proverb tells us, 'Turkey, heresy, hops and beer came into England all in one year'

There are two species of turkey, the common and the ocellated; domesticated varieties bred from these include the American Mammoth Bronze, the Norfolk Black and the White. Hens lay up to 100 eggs each year; not naturally good mothers, the eggs are usually placed in incubators by commercial breeders. The greatest dressed weight recorded for a turkey is 82 lb 14 oz (37.5 kg), for a stag bird, reared by British United Turkeys of Chester. It was bought for £4,200. Stags of this size are unable to mate.

The Buff stag in the Farmyard at Chatsworth

TURKEY MEAT

The traditional time to eat turkey is at Christmas, and Americans have turkey at their Thanksgiving celebrations in November. In the last 20 years it has become popular throughout the year, being lean and economic to produce.

Wild Turkeys

The wild species is larger and more ornate than the domesticated bird. Those in Mexico have no 'beard' (the name given to the folds of deep pink skin under the turkey's beak); they have 2 more pieces of skin called 'leaders' and bright plumage with spots on the tail feathers. These feathers are often described as 'eyed' or 'ocellated'.

FACT FILE: TURKEY

Male: stag

Female: hen

Young: poult, chick

Hatching time: 28 days

INTENSIVE REARING

Turkey raising has become intensive and industrialised and most farms rear hybrids that have been developed for their meat-producing ability. Eggs are placed in incubators and hatch after 28 days. The chicks are called poults and are kept under a lamp for 4 or 5 weeks until they can do without artificial warmth. The females are slaughtered at 3 months, weighing about 4 kg and the males at 4 months, weighing 7 kg. The world's largest turkey farm is that of Bernard Matthews plc in Norfolk with 2,600 staff tending 8,500,000 birds.

White Turkeys

These are reared on commercial turkey farms today, because they are quick-growing and produce meat that is acceptable at Christmas time. Each strain is allocated a number, and those we have in Chatsworth Farmyard are called BUT 8.

Norfolk Black

The Norfolk Black is one of our oldest British breeds: it is a direct descendant of the first turkey imported to Britain by William Strickland (*see opposite*) and a good example of a breed that has been developed through the introduction of wild blood. It has wonderful purple-black feathers and a black beak. This turkey has proved a good table bird.

uff

is breed is thought to have originated from the American Mammoth
onze, the largest of the domesticated varieties. There is a splendid
ample of a Buff stag turkey in the Chatsworth Farmyard; it has deep,
olden plumage, a splendid 'beard' and 'leader'.

URKEY EGGS

he turkey's egg is larger
nan the hen's but has
lmost the same mild taste.
: is brownish in colour and
potted.

HOW THE TURKEY GOT ITS NAME

The turkey first arrived in Europe in about
1523-4, brought by Levant (Turkey)
merchants. The English were unable to
pronounce its Mexican name - uexolotl - and
so they called it the 'turkie cock'. Archbishop
Cranmer is known to have eaten turkey, and
William Strickland, credited with
introducing it to England, incorporated the
'turkey-cock in his pride proper' into his
heraldic arms, granted in 1550.

Saanen

The Saanen, named after the Swiss valley where it comes from, is perhaps the best dairy breed. Its milk is much easier to digest than cows' milk, which makes it a perfect alternative for people who have dietary problems, suffer from asthma or have an allergy to cow's milk. The British Saanen is white and bigger than other breeds.

Cashmere

Cashmere is not a separate breed of goat; all goats, except Angoras produce cashmere – which is the soft 'down' undercoat of the goat. Some produce cashmere when young. The production of cashmere in quantity and length of hair varies depending on the climate they originate from – China, India, Mongolia, etc.

Angor

Originally from Turkey, this goat is famous for it long, silky wool. The fleece of the Angora i known as mohair, which is Arabic for 'goats' hai material'. The breed first became popular in Africa and in the mid 19th century the European recognised the potential of its fleece; th Americans soon followed sui The Angora is smaller than most goats; it ha long, floppy ears and both males and females hav long horns. Angora wool comes from rabbits

Anglo-Nubian

It is thought that the Nubian goat originated in the Nubian desert in Africa, but it has also been established in India and the Middle East for many years. During the 19th century it was crossed with English breeds - hence the Anglo-Nubian with its short coat, long, floppy ears and the 'Roman' nose which gives it a haughty appearance. The milk of the Anglo-Nubian contains more butterfat than that of any other breed.

British Alpine

Perhaps the most striking of all goats, this breed is tall and elegant with a short, sleek, black coat with white markings. It produces large quantities of milk and does well in free-range conditions. If restrained by fencing, it is more than ready to demonstrate its ability as a jumper!

...oggenburg

...long with the ...anen, the ...oggenburg is one ...the best known of the ...wiss goats, and many of ...e breeds found in Europe ...day are closely related ...it. It is light brown in ...lour and has horns.

Goats

Although not common in Britain, more people in the world rely on goats than on cows. They are commonest in hot, dry countries where there is insufficient grass for cattle. They hate rain. There is a large number of breeds of goat, and they are usually distinguished by whether or not they have pricked-up ears, and by the length of their coat.

The original English goat was small with short legs and could be various shades of grey. It is almost extinct now because of cross-breeding. Most modern breeds originated in Switzerland, especially those found in England, France, Germany and Scandinavia (as well as Switzerland). In Malta, goats' milk is an important part of the people's diet. They are also found in large numbers in northern Africa, Spain, Italy, Greece, Israel and Syria: here they are usually black, sometimes with white spots, and have lop ears. Indian goats are also lop-eared.

Breeding and raising goats

Goats are by nature browsers rather than grazers. They prefer leaves and shoots to lush grass and so clear the undergrowth of a wood. By standing on their hind legs they can reach up to branches of tre

When kept as milkers, diet is important. Goats ar most particular about the freshness of their food ar will not eat hay or other food that has been trodden on, so their ration must be put in a rack off the ground.

GOATS' MILK

Goats continue in milk for a longer period than cows - up to 2 years afte they have had a kid - and some produce as much as 5 litres a day. Mrs Clarke, who lives on the Chatsworth estate told me of a 5-year old goat which had *never* had a kid yet she gave 3.5 litres of milk a day. Goats' m is extremely important to those allergic to cows' milk and its various products. According to an old Gaelic saying, it was used as a cosmetic, too:

'With violets and the milk of goats anoint thy face freely,
And every king's son in the world will be after thee, my dearie.'

GOAT PRODUCTS

In addition to the fleeces, skins (*see page 19*) and dairy products, goats' meat is very popular in Greece, Italy, Spain and France. Kid meat tastes a bit like lamb but has a more delicate flavour.

Adult billy goats produce a powerful smell from the glands behind their ears and it has been written, 'there is no creature that smelleth more strongly as doth a male Goat . . . Tiberius Caesar, who was such a filthy and greasy-smelling old man was called an Old Goat.'

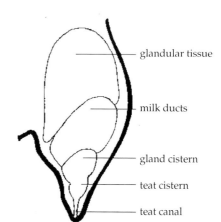

glandular tissue

milk ducts

gland cistern

teat cistern

teat canal

UDDERS

The udder of the nan goat is round, firm an silky and the teats are tapered, pointing forwards. For hand milking, a teat is held each hand and press applied by the fingers to expel the milk. The pressure is then released to allow the teat to fill with milk again.

AGGRESSION

Billy goats, like bulls, boars and stallions, can be fierce and are quite likely to charge, with head down and horns forward, at another animal or human.

HERDS OF GOATS

Herds of goats are, surprisingly, rarely seen in Britain given the quality and quantity of milk they produce. This herd is unusual in that the goats have not been dis-budded.

̶EETH

̶ goat has 6 teeth called molars on ̶ther side of its upper jaw, and 8 ̶lled incisors on the lower jaw. ̶here is a hard pad on the upper ̶w. During its first 5 years of life ̶e goat's milk teeth are replaced by ̶ermanent ones and from then on, ̶e age of the goat can be gauged by ̶e condition of its teeth.

GOATS' HORNS

Goats' horns are spiral or scimitar-shaped, which means that, unlike sheep's horns, they never grow outwards. The base of Swiss goats' horns is broader than that of the Asian goat. It used to be believed that goats' horns placed under the pillow cured insomnia and that, when the horns were burned, the ashes drove away serpents. To avoid damaging each other most goats are 'dis-budded' at a few days old. This prevents their horns from growing.

– trim

FOOT TRIMMING

In its natural habitat, a goat's hoof would be worn away by craggy rocks or gritty sand. But nowadays most goats are kept on soft turf instead of hard ground, so their hooves grow too long.

YOGHURT

With the trend towards 'natural' food, yoghurt made from goats' milk has become popular, especially the natural yoghurt from Greece which is made from whole (unskimmed) milk.

̶HEESE

̶oats' milk and the cheese made from it have a distinctive flavour. But, like ̶e yoghurt, cheese can be bought without additives, and it is also ̶nportant to those needing an alternative to cheese made from cows' milk. ̶f course, there are those who love the strong flavour.

ANTLERS

Antlers are like branches of trees with sharp tips called points. They are made of bone and are shed each spring (the horns of other animals such as sheep and cows are made of material more like nails) and it takes 4 months for the new antlers, covered in velvety skin, to grow. During this time the stag is described as being 'in velvet'. When the antlers have fully developed they begin to itch, then the stag rubs them against trees to clean the velvet off, leaving his points shining white. An extra point is normally grown each year. Mature stags have up to 12 points, when they are called royals.

The red deer stag is called a knobber in his first year when he has two knobs; at five he is a full grown stag. A royal is one with 12 points or more. With 14 he is an imperial. The few stags that fail to grow horns are known as hummels.

The antlers of the fallow deer are flatter than those of the red deer and when the bucks are mature they can be very elaborate.

The roe deer's are short and never exceed 6 points, with only two branches.

The muntjac has short, sharp antlers with few or no branches.

Deer

There are over 40 different species of deer around the world and 6 of these may be found living in the wild in Britain; red, fallow, roe, Japanese sika, muntjac and Chinese water deer. The latter 3 breeds have escaped from parks. They are cloven-hoofed herbivores and feed off every sort of plant, even ivy and yew that are posionous to other animals. Like cows, they are ruminants, and have 4 stomachs (*see page 11*). In all species except the reindeer, only the male deer - the stag - has horns which are shed every year in the spring and called antlers.

Since the Middle Ages, deer have been hunted for sport and kings have protected them. The red deer stag was considered the king's property; only he and his guests were allowed to hunt it. The meat is called venison and has a very rich taste.

Deer mate in autumn. During this 'rutting season' the stags and bucks fight to win the hinds and does and can be dangerous to humans as their instinct is to possess their females and drive rivals away. The stag 'roars' a challenge to his rivals when rounding up his hinds.

They can jump very high so the deer fences are extremely important. They can swim even when very young so the river on the edge of the Park at Chatsworth is no barrier to them.

Red Deer

The red deer has lived in Britain for thousands of years. Its natural home is open forest but, due to the felling of forests over the years and because it is a destroyer of crops, it has been driven to the moors, mountains and woodlands. It has a reddish-brown summer coat that turns greyish and thick in winter. Stags live in groups apart from the hinds for part of the year. In the rutting season they fight for the hinds and the strongest wins the most wives. The herds spend the winter and spring together until about June when the hinds like to be alone to have their calves, hiding them in bracken or rough grass.

The stags do not use their points when fighting, but the antlers interlock and they push hard against each other. Their strength is in their weight and the loser is seldom hurt. The stag needs to eat lots of calcium and other minerals in order to grow his antlers every year.

Muntjac

The muntjac deer, known as the barking deer, is smaller that the roe deer, only about 50 cm high. It has 2 teeth that grow in to tusks which it use when fighting. It originated in China and India and its reddish-brown coat serves as an efficient camouflage in its habitat in the woods of the south of England. The muntjac lives alone or in pairs. The rutting season varies and the young are born during the summer.

Deer in Chatsworth Park

200 fallow deer and 150 red deer roam Chatsworth Park. During the winter months when grass is scarce they are given the same extra food as the sheep in the form of silage, hay and concentrates. The rutting season is in October and November, and the young fallow (fawns) and red deer (calves) are born in June. Culling (when a certain number is shot) takes place in September and the meat (venison) is sold in the farm shop. The deer do not need much attention and keep themselves well hidden most of the time in the 80 hectares of the Chatsworth Park that are reserved for them and not open to the public.

Roe Deer

The attractive little roe deer, known as the fairy of the woods, has lived in Britain since pre-historic times, but at one time it nearly became extinct. However, it is now common in some parts of the country. It is very shy and has even been known to crawl on its knees to evade enemies. It has a white tail and its fox-red coat is replaced by a brown coat in the winter. The roe differs from the red and fallow deer in that it lives in family groups consisting of a buck, a doe and two fawns (they usually have twins). The young are speckled with white spots. The rut takes place in July; the roe buck may have more than one wife but not as many as red and fallow deer. Although they mate in July, the babies do not begin to develop inside the mother until about December. This is a strange breeding cycle known as 'delayed implantation'.

Cloven-Hoof

Deer chew the cud and are cloven-hoofed, like cows, goats and sheep; their horny hooves are split in the middle. It is worth examining muddy places for their footmarks, called slots; young deer leave slots close together, the adults' deeper prints are further apart.

Fallow Deer

Smaller than the red deer, the gentle fallow deer can be seen throughout Britain. It is reddy-brown or yellowy-brown and the white spots along its back are excellent camouflage. The fallow deer originated around the Mediterranean and today many live wild having escaped from parks. When disturbed its white tail stands up, showing a white patch which warns the rest of the herd as it runs away.

During the rutting season the buck marks out an area by pawing with his hooves and sweeping his antlers through surrounding bushes and branches, leaving a scent on them that comes from special glands under his eyes. He walks up and down this patch, roaring, and the does come to him. If another buck appears, they will fight. With heads lowered they push against each other and the stronger wins the doe.

Japanese Sika

Similar in behaviour to the red deer, although smaller, the Japanese sika deer, imported to Britain about 350 years ago, has a dark brown coat that is spotted white in the winter. It has smaller antlers and makes a whistling noise.

FACT FILE: DEER

Male: *buck, stag*

Female: *doe, hind*

Young: *calf, fawn*

Gestation period: 230 days

Cock Pheasant

The common pheasants that are reared or released on most farms and estates, like Chatsworth, are the Eastern Chinese Ring-necked Pheasant, the Blackneck Pheasant and Melanistic Pheasant. The cock is a pugnacious bird. He is handsome with a brilliant plumage and red wattles. (The wattle is the brightly coloured piece of skin that grows on the bird's head or neck.) The wings are short and the tail long and both the cock and the hen run and fly well. In courtship, the adornments of the cock are strikingly displayed.

Pheasants and Partridges

Pheasants are extinct in their natural habitat of the Middle East and China. Pheasants and partridges are game birds, a term which describes birds and animals hunted for sport and prized as food. Others include grouse, quail, duck, snipe and mammals such as deer and hares. We rear pheasants at Chatsworth and we also have a few partridges and ducks.

Partridge and quail belong to the same family as the pheasant. Some breeds are kept purely for ornamental purposes and the Golden, Silver and Lady Amherst pheasants may be seen in the aviary in the Chatsworth Farmyard.

BITTING

When large numbers of pheasants are kept together in enclosed areas, they tend to peck each other's feathers. So they have small metal rings fitted between their upper and lower beaks to prevent them from closing their beaks tightly. This means they are unable to get a good hold of the feathers but at the same time they can eat and drink normally.

REARING PEN

Chicks are kept in a brooder with a lamp to keep them warm until their feathers have grown at about 6 weeks old. They have a run so they can go outside. The poult is then put into a release pen in the woods where it is still protected but it is free to fly and roost in the branches of trees. The pen is open to the sky; most poults return to the safety of the pen at night.

Partridge

The most common partridge is the grey *Perdrix perdrix*. The red-legged or French variety, and we rear a few of these on the estate at Chatsworth. It is also found in Europe. They pair early in the year, nest in fields and hedgerows and each hen lays about 15 olive brown eggs. The cock attends the hen while she is sitting and the young stay with the parents for several months, forming coveys, the name given to a group of game birds, of about 20 birds

The pheasant egg is of varying shades of fawn and olive green. The egg is placed in an incubator, where it takes 23 days to hatch. When it is ready to hatch, the chick chips its way through the egg shell with its beak.

THE KEEPER

The keeper starts work as an under-keeper and he gains his experience from working with the birds. An important part of his job is to control the vermin - animals that are a threat to the protected game and their habitat. Rabbits, squirrels and foxes are considered vermin, and may attack and kill young chicks. The head keeper at Chatsworth supervises the game on the estate, including the deer. He and his colleagues are the eyes and ears of the remote places on the estate.

GUN DOGS

The Labrador retriever was brought over from Labrador, in north eastern Canada, over a century ago and is now the most popular of all retrievers, being strong and easily trained. It fetches the fallen game and brings it back to its handler. It is also widely used as a guide dog for the blind. The English springer spaniel is the most popular of the sporting spaniels, particularly for rough shooting, and it is much used by keepers and beaters for flushing out the birds.

Farm Wildlife

Wild animals live in our farming countryside and farmers and landowners are aware of the necessity to protect their natural habitat since many of the species have become endangered due to the destruction of hedges, trees, barns, rivers, ponds, streams, fens and marshes. Otters need ponds, streams and rivers; hedgehogs, rabbits, hares, harvest mice, dormice and shrews need hedgerows and fields; squirrels, badgers, fox, deer and wild cats need woodland.

Pigeon
The wild pigeon, developed from the dove and has become a real pest to the arable farmer. swooping down on newly-sown fields to eat the seeds. Farmers install bird-scarers (replacing the 'scarecrow' in their fields), machines that let off loud bangs every few minutes, t frighten the pigeons away.

Badger
Badgers are one of the biggest wild British mammals. A boar can weigh up to 18 kilograms. It is a nocturnal creature. It lives in a hole called a sett, and succeeding generations make new nurseries for the 2 or 3 silver-grey cubs born in spring. Despite the phrase 'smelling like a badger' they keep their setts clean. They carry bracken and leaves to the mouth of the sett and work the bundle backwards into the hole.

Badgers live on rabbits, grubs, insects, beetles, frogs, worms and fruit. They love honey and will eat bees and wasps' nests. They sleep a lot in the winter, but do not truly hibernate as they come out on mild nights to look for food.

Many farm animals such as the badger and the fox, being nocturnal, will usually be seen only at dawn or dusk.

Rat
The common or brown rat is Public Enemy No 1. They are very cunning and difficult to catch. Rats will eat anything from the best cereal crops to soap and candles. They are destructive, savage and carriers of disease. They do no good to anyone and have no redeeming features.

Rabbit
Rabbits devour crops and, in a snowy winter, they destroy newly-planted trees by gnawing the bark right round the stem. They live in holes called burrows and many burrows together are a rabbit warren. Rabbits have big families so their numbers increase very quickly.

In the 1950s a virus called myxomatosis was introduced. This killed a huge number of rabbits, but a few survived and although the disease still breaks out from time to time, many are immune to it and so the rabbit population is increasing again.

The Harvest Mouse
The harvest mouse (*mus minutus*) is the smallest of our mammals except for the pygmy shrew. It has brown fur, white underneath, and its very long tail is prehensile (can hold on).

It makes a perfect little nest round the stems of standing corn. The nest has no opening and the mouse push its nose in anywhere it chooses and squeezes its body in. In the winter they nest in a hole in a bank or a stac of straw bales.

The Dormouse
The dormouse is the only rodent which hibernates. It makes a nest in the bottom of a hedgerow with wall and roof of moss, honeysuckle bark and dead leaves, lined with feathery grasses. It eats all the nuts and berries it can manage in the autumn and then packs itself into its nest an sleeps till the spring.

Hare
The common, or brown, hare and the mountain, or blue, hare are related to the rabbit but they are larger and have long ears with black tips.

The males are called bucks or jacks, the females does or jills, and the young are leverets. The leverets are born in a 'form', a shallow hollow in the ground in long grass or a ploughed furrow.

They do not live in burrows, like rabbits, but they are equally at home on pastures or cultivated corn land. The mountain hare is found on the high ground in these islands. In the winter its coat turns white so it cannot be seen in the snow.

The hare has a keen sense of smell and hearing, and its long, strong hind legs enable it to run faster than any British animal and thus escape its enemies. In the spring the jacks fight each other, sitting on their hind legs and 'boxing', with their forelegs.

The Shire

A descendant of the Great Horse of the Middle Ages that was bred for war to carry heavy armour, the Shire had enormous strength and so became invaluable for agricultural, transport and industrial work. Big, strong horses were needed in large numbers for the draining of the fens in East Anglia and Shires were bred extensively for this purpose. The Shire Horse Society was formed in 1884. Although by the end of the Second World War there was little use for the Shire, the breed has survived and is often seen being used to deliver beer to public houses in towns and at many county shows throughout the country.

The Percheron

The Percheron originated in France and is found all over the world. The English breed society was founded in 1918 when an importation from France arrived here. Percherons are always black or grey, often strikingly dappled. Like the other heavy draught horses, they came perilously near to extinction in the 1960s but are enjoying a revival now.

Farm Horses

Horses are described by their age: as a foal, yearling, 2 year old, 3 year old and so on till 8 years, after which they are 'aged' – that is their age is taken from 1st January of any year, irrespective of which month they were born. Horses are measured in 'hands', from the ground to the point of the withers (shoulder). A hand is 4 inches.

Cart horses gradually took the place of oxen as draught animals on farms, pulling ploughs, harrows and other farm implements as well as wagons and carts. A million heavy horses were working in England and Wales in 1913.

A Shire stallion can weigh more than a tonne and often stands 18 hands high. They are docile and easy to break in to harness; the 'gentle giants' of the horse world.

Successive Dukes of Devonshire kept a stud of Shire horses and so the tenant farmers and his other neighbours had access to the best stallions.

FACT FILE: HORSE

Male: *stallion or Entire* – kept to mate with mares
gelding – castrated male
colt – until the age of 4 years

Female: *mare*
filly – until the age of 4 years

Young: *foal*

Gestation period: 340 days

Rare Breeds

It is now universally accepted that a threatened species of mammal and bird, whether it be wild or domesticated, must be preserved. Often the original species has become rare and sometimes even extinct, not through neglect or slaughter but through the human urge to 'improve' the breed for utilitarian purposes. The consequent inter-breeding and cross-breeding has introduced risks to good mothering, resistance to disease and hardiness. So it is of great importance to maintain 'primitive' varieties that have evolved naturally. Specialised breeding of farm stock started only about 200 years ago.

Bagot Goat

The handsome Bagot goat has a white body and distinctive black head, neck and shoulders, a long shaggy coat and horns that grow straight upwards curving slightly backwards. It is said that Richard the Lionheart (1157-1199) originally introduced the breed to England after a crusade. It has since lived semi-wild in Bagot Park, Staffordshire, purely as an ornament, for the Bagot goat has little to offer; the ewes produce very little milk, have difficulty in producing young and are bad mothers. However legend has it that when the goat at Bagot disappears, so too will the family who live there.

White Park Cattle

For over 700 years the Chillingham herd, the best-known of the White Park cattle in Britain, has been pure-bred (without the introduction of foreign blood). It is said that the herd originated from the wild cattle which roamed the Caledon Forest until Chillingham Park, Northumberland was enclosed and they have remained there since. The same story applies to those herds now in Cadzow Park in Scotland, Chartley Park in Staffordshire and Dynevor Castle in Wales. They are white with black tipped horns, muzzles, eyes, lashes and hooves, and their horns curve upwards and slightly inwards at the end. White cattle have always had a certain mysticism about them and have played an important part in religious ceremonies in many cultures.

Derbyshire Redcap Fowl

The Derbyshire Redcap – so called because of its red comb that is broad and pointed looking like a cap – is one of our oldest breeds of fowl and can be traced back to the 14th century in Derbyshire and Yorkshire. Although it is a small breed it has a large breast and produces good meat; it is a good layer as well. The hen has brown feathers with black tips, grey legs and red eyes; the cock has golden feathers on top with golden feathers edged with black on its hackels, wings and underneath; on the tail and wings it has a strip of very dark silver/green feathers. The hen's eggs are white with a slight tint and have a beautifully delicate flavour. The hen rarely goes broody.

Exmoor Pony

The Exmoor pony is the oldest British pony breed and can be traced back to pre-historic days. It is a small, hardy animal that can survive the bad winters with little shelter on the moor. Generally brown with black points, it has a mealy nose, sharp ears and short legs.

Large Black Pig

The Large Black pig was developed in the 19th century in East Anglia, Devon and Cornwall and was also found in South Africa and South America. It is large and black with lop-ears and used to produce good bacon due to its long back. Before the white pig became popular for its carcass, the 'blue' pig (a cross between a Large White boar and a Large Black sow) was much in demand. This is a quiet and docile animal that can tolerate cold and hot climates. The sow is an excellent mother who rears strong healthy piglets.

Whitefaced Woodland or Penistone Sheep

The Whitefaced Woodland sheep comes from Woodland Dale, a valley in the Pennines between Derbyshire and Yorkshire, and nearly 90% of the breed are found in that area today. They are sometimes referred to as the 'Penistone' after the name of the town where they used to be marketed. It is one of the largest hill sheep (not as hardy as the mountain sheep), long-legged with a white fleece, face and legs and a pink nose. Both the ewe and ram have horns.

RARE BREEDS SURVIVAL TRUST

The Rare Breeds Survival Trust, which was started in 1943, has done a great deal to save many breeds of farm animals and birds on the verge of extinction. Now there are a number of farms, open to the public, which specialise in rare breeds. A list can be obtained from the Trust at the National Agricultural Centre, Kenilworth, Warwickshire, CV8 2LG.

Suffolk Horse

The Suffolk can be traced back to the 16th century. It is always chestnut in colour, ranging from red, gold, copper, through to dark and light liver, with no white markings and is 'clean-legged' (does not have the feather or long hair below the knee of the Shire horse). This suits the heavy clay on which it works. It is very hardy and has a particularly gentle nature.

Cereal Crops

Cereal crops are varieties of grasses (*Gramineae*) that are harvested for their seed (grain). 7 main varieties of cereal are grown throughout the world; barley, maize, oats, rice, rye, sorghum (known as Indian millet) and wheat. Barley, oats, rye and wheat are grown in Britain. Wheat was the main crop until the Romans introduced other varieties from eastern Europe in the 1st century AD, together with new methods of cultivation. The cultivation of cereals has been made much easier and quicker with the advent of machinery. Originally an ox pulling a plough with one blade was used to till the land and the seeds were broadcast by hand. Harvesting and threshing were done by hand with scythes and flails, a scythe being a long, curved blade with a wooden handle, and a flail a wooden pole attached to a wooden handle in such a way that it swings and beats the seed out of the seed head. Today wheat and barley are the most important cereals grown in Britain. Some varieties are sown in autumn (known as winter wheat or winter barley) and some in spring; they are harvested from early July onwards.

BURNING STRAW OR STUBBLE

In corn growing areas where few anima are kept, it is sometimes easier and cheaper to burn straw (stubble), as balir and transporting it to farms as bedding and fodder for animals is expensive. Th has been a controversial issue for some years since people are concerned about polluting the atmosphere. Straw burnin will be against the law after 1992. Various chopping machines have been invented, enabling the surplus straw to be ploughed into the land after harvesting or made into briquettes to fu domestic boilers.

CORN DOLLIES

The last sheaf of wheat standing, or straw from the last cartload carried home was woven into corn dollies and Christian crosses, used as offerings in churches at Harvest Festival. In the olden days these corn dollies had magical attributes said t placate the goddess of the harvest and to aid fertility.

The Shandy Barrow
In the early part of this century, before the advent of mechanisation, fields were sown either by hand or from an implement such as the shandy barrow.

Oats
The land devoted to oats has decreased since barley was introduced and proved a more valuable crop. Oats are still greatly valued as a breakfast cereal and in oat cakes, particularly in the North of England and Scotland. Rolled oats are fed to horses. Oat straw is a valuable winter feed for cattle.

Barley
Barley has long, beard-like spikes called 'awns'. When it is ripe the ears hang down. Barley is used as animal feed and to make a small amount of human food (pearl barley and breakfast cereals). It is also malted – when the grain is germinated and turned into 'maltose' sugar which is used to produce alcohol in beer and whisky. The grain sold to 'maltsters' fetches higher prices than feed barley.

Wheat
For bread making it is necessary to use 'hard wheat'. Not all the wheat grown in this country is of this type as hard wheat does better in a warmer climate. Some is imported from Canada and America to add to ours for bread making. At Chatsworth we grow, and mill into flour, enough to make the bread we sell in the Farm Shop. The rest is sold to be made into breakfast cereals and biscuits and food for livestock.

COMPUTERS IN FARMING

Computers have become an important part of farming; in the farm office they keep records of the accounts, track each animal's progress and keep stock reports; they detail the jobs done and requiring to be done, enabling the farmer to be more accurate in forecasting profitability.

The cows in some dairy herds are fitted with electronic collars, which are connected to a computer. The cow's milk yield is weighed and recorded and the cow fed concentrates according to that yield. The transponder in the collar conveys the information to the computer, which can be used to keep records on individual cows.

Many tractors and combine harvesters are also fitted with small computers enabling the driver to check with a central system that his details are correct. Tractors today can have as many as 26 gears and need the support of such computers for maximum efficiency.

Combine Harvesters
Combine Harvesters are the most expensive machines on a farm and cost up to £100,000. Claas (German), New Holland (American) and John Deere (American) are the most frequently seen.

Grain

Cereal plants have a head (the ear) containing grain that is the food product. The process of harvesting cuts the crop and separates the grain from the stalk, which is left as straw and used as fodder and bedding for animals. When grain leaves the farm, it should contain no more than 14½% moisture. In a dry season it will be harvested at this level or below but in a wet season the moisture content of the corn may be as high as 25 – 30%; then it is dried.

Maize

Maize came from South America. In the southern part of England it can be grown as sweetcorn (corn on the cob) because the climate is warm enough; elsewhere it is grown as excellent fodder containing 10% protein. The whole plant can be harvested green and made into silage.

Farm Buildings

The farm as a landholding developed as an economic concept in the Middle Ages. The farm buildings on those landholdings are now very much a part of our social history and reflect the prosperity (or lack of it) of the times. Some are built of timber, some of stone and some of brick; they show the use of local materials and also reflect the local styles of architecture. There are few of these buildings left.

From the time of the Agricultural Revolution in the 18th century there have been farmyards – that is barns, cow sheds, stockyards and pigsties arranged formally. Some were U- or E-shaped, others built in a square. Today, on large farmsteads where bulk production is of the essence, animals are kept in purpose-built buildings, and fed on grain stored in huge silos, which sometimes look out of place next to old barns. The problem of beautiful farm buildings which are of little use to the farmer is a worrying one for lovers of British rural scenery.

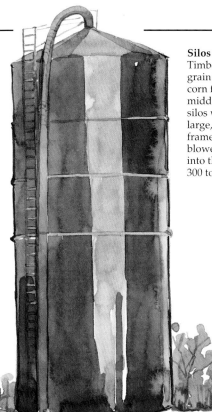

Silos

Timber-framed barns with grain bins were used to store corn for winter feed until the middle of this century when silos were introduced. These large, round, galvanised steel framed and domed silos have blowers which suck the grain into the silo. Large ones hold 300 tonnes or more of grain.

The Cow Shed

From April till October most dairy cows and beef cattle live out in the fields. In the winter they are usually housed in large barns. The modern ones are built of timber or steel frames with pillars and corrugated asbestos roofs. An 'umbrella' shed is walls and a roof where the interior can be adapted to any use from housing stock to storing corn. It is usually open on one side for ventilation and will have plentiful water supplies, wire and steel racking for hay, silage and concentrates, and a salt lick to provide essential minerals. The concrete floor is covered with straw or woodchip. The dung from the cattle is trodden into the bedding and the manure is spread on the fields as fertiliser to give back to the land some of the goodness which has been taken out of it.

The Hay Barn

Hay, grain and livestock were often housed in one building in the old days. With increasing mechanisation in the early 19th century, coinciding with higher crop yields, there was a need for farm buildings with flexibility. The Dutch barn, from Holland, was widely introduced. Originally it was built of wood pillars with a rounded wooden roof; then it had iron posts and a corrugated iron roof. Today, with the enormous round bales being stored in polythene sacks, the traditional hay barn is often used for other purposes, such as winter shelter for young cattle or lambing pens.

Hedges

In lowland areas where there are plenty of bushes and trees, the fields are enclosed by hedges. Whitethorn (quickthorn or may) is the best hedging plant. A cut and laid hedge requires special skill and a well-made one lasts for 15 to 20 years. Stakes are driven into the ground 1 or 2 to the metre, and the existing bushes are 'layered' or cut at an angle and bent over to weave, layer upon layer either side of the stakes. They are not cut through so regrowth is encouraged, and they sprout in the spring. The top of the hedge is finished with 'binders' wound between the stakes. Hazel and ash are good because they bend without breaking.

Dry Stone Walls

Around Chatsworth, and in many upland areas where there are few trees and bushes but stone is plentiful, the fields are enclosed by walls. These walls, as their name suggests, have no mortar or cement to hold them together but are built by hand, stone upon stone. The bigger stones form the sides and the cavity in the middle is packed with small stones called 'fillers' which prevent the big ones from collapsing inwards. Every so often a 'through stone' – big enough to reach from side to side – is placed across the wall. The 'throughs' hold the sides together, preventing bulging and strengthen the edifice. The tops are finished with 'coping' stones which are set at an angle to those forming the wall itself.

Stone walling is a highly skilled job and cannot be done in a hurry. Competitions are held regularly by The Dry Stone Walling Association of Great Britain in stone wall country where you can watch the experts and the learners building walls.

Gates

Field gates are 5-barred and made of wood. Some are made of steel, and dipped into a galvanising fluid to prevent rusting, with either 6 or 7 bars. The gate is hung on two posts, which are usually made of wood. The width of gates varies.

Barbed Wire

This is the most commonly used fencing on farms. It comprises twisted wire strands, with short, sharp prongs every few centimetres, which snag animals and prevent them trying to push their way through the wire and the fence. It is nasty but effective on cattle or sheep, and they are rarely hurt. It is no good for horses because they get cut very easily.

Fields

After the decimation of Britain's population by the Black Death in 1384 - 9, leading to labour shortage on the land, the idea of an enclosed 'field' became common. However, most of the field patterns we see today are the result of a series of Acts of Parliament in the 18th and 19th centuries.

The enclosed fields encouraged the landowner to improve the land, thereby making it possible to increase the numbers of livestock. Boundaries were made of whatever suitable material was near at hand.

Hedgerows

The hedgerow is an important habitat for game birds, song birds, small mammals and a wealth of wild flowers.

Farm Machinery

Machines have replaced horses as the source of power on the land in Britain and are used in all farming activities. They save time, reduce the number of employees and increase productivity. However, in third world countries such as many parts of Africa labourers, often women, still do most of the work although machines are gradually being used more and more.

There are machines for preparing the soil, helping the crops to grow and harvesting them; for trimming hedges, mowing and turning hay in the field, gathering it and making it into bales; mowing and carting silage; drying grain; milking and feeding. The first mechanical reaper (or harvester) was invented by a Scottish church minister in 1826, and was pushed by horses. Then came the threshing machine driven by a steam engine. Threshing is separating grain from the straw after harvesting it. This used to be done by men using flails to beat out the grain from the seed heads. A machine was invented to 'combine' the reaping and threshing machines but it took forty horses to pull it which was not practical. The modern combine harvester works on the same principle.

Early this century the tractor was introduced to take the place of the heavy horse to do all the pulling, pushing, lifting, loading and carrying on the farm. The first British tractor was built in 1903, but horses were still used all over the country till after the Second World War, and not till the late 1950s did tractors finally replace them.

Tractor

The tractor has a very powerful engine and large wheels and tyres. Most tractors have 4-wheel drive, which means that the engine drives all the wheels instead of just 2 as in most cars, giving it a better grip on wet, muddy and slippery ground, especially on slopes. The tractor performs two very useful functions: power take-off - which means the running of other machines such as mowers, drills, broadcasters and balers by connecting them to the tractor's engine, and hydraulic lift - lifting and lowering heavy machines attached to the tractor, enabling the farmer to lift a plough off the ground, for example, carry it the field and then lower it into place.

Trailer with Grain

When the grain tank in the combine harvester is full, the light on top flashes and a tractor pulling a trailer drives up alongside the combine. The auger on the combine (the arm through which the grain passes) swings over the trailer. They drive along the field in convoy, the combine continuing its work, the grain is levered up through the auger into the trailer. When the grain tank is empty, the auger swings back to the combine and the tractor takes the grain to the farm buildings to be dried and stored.

Fertiliser

During the preparation of the soil, fertilisers are added to improve the growth of the crops. Farmyard manure is the natural fertiliser, put on the fields in liquid or raw form. This coating fertiliser distributes chemicals and can add essential minerals to balance the composition of the soil.

Drill

The drill plants the seeds in the ground to an exact depth and in neat straight lines so that the farmer can weed in between the rows. It has two containers; one for seeds and another for fertiliser which is added at the same time. The drills can be set to sow seeds of all sizes from tiny grass seeds to the larger peas and beans. Some cut channels in the ground, drop the seeds in and then cover them over with soil in one operation; precision drills plant the seeds individually. Potatoes have their own machines to plant the big tubers, or seed potatoes.

Plough

The plough, pulled behind a tractor, is the first implement used in the preparation of the soil. It has large cutting blades called ploughshares which may be reversible, so that the plough can be turned over at the end of the field to plough the next furrow. The ploughshares cut through the soil, turning it to bury the weeds and stubble and helping circulate the air and any water in the soil. They leave large chunks of earth, known as clods, and furrows that are about 25cm deep.

Harrow

When sowing small seeds, such as grass, the earth must be broken down into a fine seed bed of smooth firm soil; after ploughing, it is necessary to use a harrow or cultivator. The rotary cultivator is one and the disc harrow is another that also moves the earth from side to side as it chops it up. They are normally used immediately after ploughing.

Boom Sprayer

Crops need to be protected from weeds and diseases to enable them to grow. Herbicides kill weeds, pesticides and fungicides control diseases; and fertilisers provide nutrients. These are applied either in liquid form for spraying or in solid grains for broadcasting. The liquid is spread over the fields by a spraying machine, attached to the tractor. The sprayer has a tank containing the liquid chemicals and two long booms along which are lots of holes.

Index